Preface

Is there really a "creative phase" in science? You bet your booty there is. Diagramming it, it looks like this:

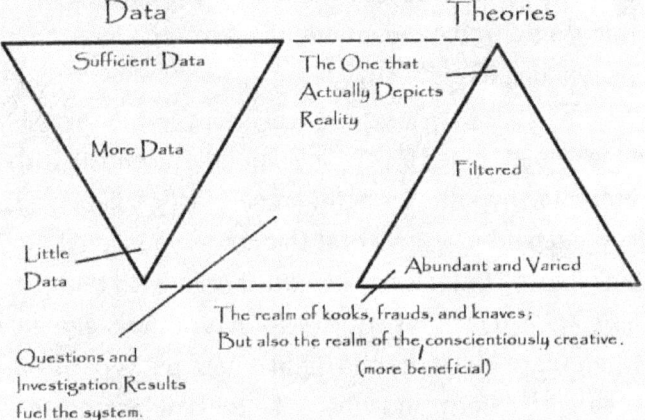

Data   Theories

Sufficient Data | The One that Actually Depicts Reality

More Data | Filtered

Little Data | Abundant and Varied

The realm of kooks, frauds, and knaves; But also the realm of the conscientiously creative. (more beneficial)

Questions and Investigation Results fuel the system.

Explaining: When there is still little data to go on (and this refers to the frontiers of science), you nevertheless are dying to use your imagination, creativity, and intuition to propose an early explanation, and this is the 'creative' phase of science - it is the proper phase for creativity, and indeed, creativity is critical in weighing the best possible path to investigate further.

As more data is gained (through hard, tedious work and brain-breaking analysis, and is cross-discipline verified through slow, lengthy processes) (making for 'good data'), then the creativity aspect diminishes - the variety of hypotheses dwindle as the weaker hypotheses tossed into the Discarded Hypotheses Bin where writers of sci-fi forage at night.

As the chart depicts, when there is finally sufficient data, then the hypothesis that actually predicted reality wins. In the meantime, there may be more than one working hypothesis.

Is that a bad thing? If you consider how scientists with competing hypotheses fight, argue, slander, libel, and go at each other's throats at this stage, you would think it was - but it is not - creativity is a good thing at this stage - in fact it is critical - for it is at this stage (when there is still so little data) where paths of future research are born, and where the tools of 'perspective' are generated - tools with which we peer into the unknown with (and the more we have in our perception toolbox, the better equipped we are to peer into the unknown).

As the chart also depicts, when there is little data, there is still room for kooks, frauds, and knaves to make claims that they already have the definitive answer (hoping they can cash-in somehow from it). Their wild claims do serves a few purposes - that of making more worthy hypotheses look that much better, and opening the minds of scientists to unpredicted possibilities.

So, in this book, I've donned my creativity hat (hopefully not tainted by 'kook', 'fraud', or 'knave'), and I've collected my creative contributions (to date) to those scientific frontiers where there is still little data to go on, and where I am not sufficiently satisfied with the current hypotheses. I also make contributions to those scientific areas that DO have sufficient data and good working theories and models already, but which I chose to ignore - giving preference to my wild imaginings.

Do I claim that any, or all, of my creative hypotheses are correct, and are the actual depictions of reality? I cannot - for remember, this is the scientific phase when there is not enough data. I can only contribute to the myriad of other possibilities, and it is up to researchers to choose which are worthy of further consideration, funding, study, exploration, testing, time, headaches, and reporting.

What do I hope will happen, then?

Best Case Scenarios
The best result is that I will identify new worthy paths of immediate inquiry. The second best case is that, while not worthy of immediate further study (or perhaps it is not possible, lacking futuristic equipment), they will be kept in the back of theorist's minds as nagging possibilities, and will be worthy tools of perception.

Worse case
They will be laughingly and immediately discarded by experts because I was unaware of existing data. Best of the worst case is at least I will have entertained them! (Worst of the worst case is I will not have even achieved that).

Evil Case
There is an evil outcome - that my creative hypotheses will be shot down by those who have their own agendas - their own creative hypotheses that really hold no more merit than mine, evil scientists who have investments of self-interest at stake rather than truth - such as undeserving funding or prestige or through knavery getting the girl and riding off into the sunset.

So the stage is set. Scientific frontiers. Little Data. A good time to let the creative, imaginative, intuitive, and inductive reasoning juices flow...

Numi *Who?*

Contents

Journal 151: Scientific Vampires and Werewolves

Introduction
There have always been scientific theorists who make claims akin to claiming that vampires and werewolves exist. String Theory is a good example, with claims of extra spatial dimensions. People who latch on to multi-dimensions (more than three) also latch on to one and two-dimensional 'worlds' - which are as absurd as vampires and werewolves...

Example: 'Worlds' with Less Than 3 Dimensions
Let's compare these to "vampires".

They, like vampires, are mental constructs that do not exist in the 'physical world'. What about mathematical models and equations based on one and two dimensions, you might argue? Even vampires can be mathematically modeled, but, like one and two-dimensional 'worlds', such models will be models of 'ideas' only.

People who try to give analogies of one and two-dimensional 'worlds' give self-contradictory analogies - for they invariably use 3D object and try to have us imagine 3D entities in such worlds (such as a 3D ant crawling around an extra-dimensional loop).

Now let's seriously consider 1 and 2 Dimensions: With one dimension (for example 'length') or two dimensions (let's say 'area'), you still have 'nothing' in the 'physical' world - with no height added, you have no volume, hence no 'object' - all you have *imaginary* length (1-D) and *imaginary* area (2-D). Once you add 'height', THEN it can represent objects in the physical world. Not before - only in flawed imaginings mistaken for physical reality.

Example: A Universe with More Than 3 Dimensions
Let's compare these to "werewolves".

We have the same thing occurring here - we have 'ideas' only that do not exist in the physical world, and where any mathematical models are models of ideas.

Conclusion
We can only stay tuned to see if such multi-dimensional mathematical models like String Theory are true (and which one is true - since there are many conflicting mathematical models of String Theory, which is an assumption based on an assumption (that Quantum Entanglement works through extra dimensions), and String Theory models change as fast as you challenge them, and none are even testable -

two reasons why Richard Feynman did not like them, or take them seriously).

In the meantime, you can watch kids (and adults) mistake pure ideas for physical reality, based on people depicting scientific vampires and werewolves with flawed, self-contradictory analogies, and watch (in disbelief) as the other kids (and adults) marvel...

Journal 150: Young Mind, Old Mind

A visualization of a conversation between a young mind and an old mind:

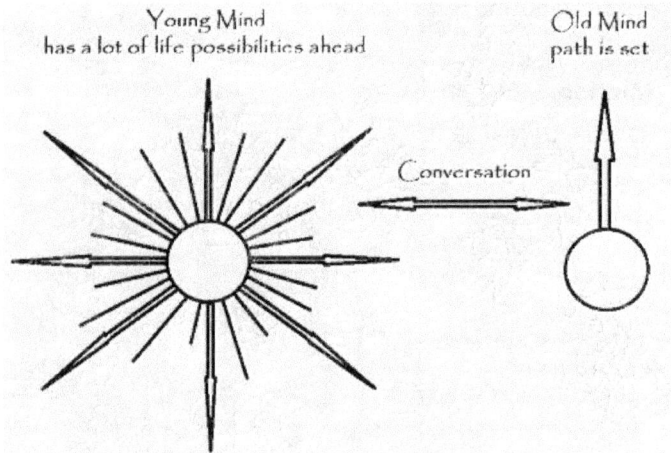

Problem Areas:
Advice
The old mind invariably attempts to set the young mind on one particular path - and invariably fails - because the young mind sees too many possibilities to try. It does not see the world like the old mind - the old mind having time remaining for but one path.

Romance

The young mind sees the old mind as but one path - and so is always looking beyond. The old mind fails to understand this.

Relationships

There will be constant friction, or there will be one unhappy party as long as this difference in outlook is not understood.

Solution:

Have both realize it.

Journal 148: Age Equalization via Time Dilation

Problem: There are two people with an age difference of 17 years - one is 59 and the other is 43. They ask you to make them the same age. How do you do it?
Solution: You keep the younger person stationary (on earth, for example) and you set the older person in motion (on a round-trip space voyage, for example), and eventually they will be the same age due to time dilation, where relative time slows as motion increases (a part of Einstein's Relativity).
How fast would the traveler have to go?

Let's put the traveler on a jumbo jet flying at the typical **600 mph**. It would take around 4 billion years for them be the same age, so the traveler has to travel a little bit faster.

Let's send the traveler on a space voyage at half the speed of light - **50% of the speed of light** - 335 million mph - which is quite fast, but which only creates a ten percent slowing of time for the traveler. The waiting person would be in a higher gravitational field (on earth vs in space), which slows time for the waiting person, which is bad; but it is negligible compared to our speed's dilation, so we can ignore it. At half the speed of light, it would take 103 of the traveler's years (119 of the waiting person) for the 17 year age gap to close, where they both would be around 162 years in age. This is much better than 4 billion years old, but it is still not practical, unless medicine and health make incredible advances on earth (and soon).

Considering this, the waiting person would be a lot healthier than the traveler at that age (since the traveler would be long dead - not having had access to the breakthroughs on earth). Perhaps the traveler might make the same breakthroughs while traveling (not much else to do), and there would be a happy, healthy reunion.

Traveling at **90% of the speed of light** halves the passage of time for the traveler, yet the traveler's trip would have to last 27.5 of the waiting person's years (12 for the traveler), where they both would be around 71 when reunited - still too old to enjoy life together, and not much hope for medical/health breakthroughs in such a short time span.

How about **95% of the speed of light**? The time dilation is 3.2 years to 1, so the traveler would have to be gone for around 22 of the waiting person's years (7 of the traveler's), so they would both be around 66 years of age - better than 71, but that's still up there in years, and a long time to wait for the waiting person...

Traveling at **99.9% of the speed of light** (a mere 669.9 million mph) is more attractive - the time dilation is a whopping 22.4 years to 1, so a journey of 16.8 years for the waiting person (only 9 months for the traveler) would do it - the traveler would return and they would be tandem surfing together in no time... and think of how wiser the waiting person would have grown in 16.8 years compared to the traveler's mere 9 months...!

How far to the next nearest star would the traveler get at 99.9% of the speed of light? The next nearest star is 24 trillion miles away, and in four and a half months the traveler would have to turn around after traveling out only 5 trillion miles, which is only roughly 1/5 the distance. The traveler would be well into the Sun's Oort Cloud, however, a place filled with rocky debris, making it a hazardous place to visit at that speed - the traveler could be vaporized...

Mass Dilation

but then again the traveler's mass would increase - but would it be enough to survive any collisions? At 99.9% of the speed of light, a 200 lb. traveler's mass would only increase to 2.5 tons (around 4500 lbs. or 2000 kg), so no - the traveler would still be unacceptably damaged - I'd set the 'overwhelming mass factor' at 100 million kg (roughly the mass of two battleships).

Length Dilation

Length shortens as speed increases. Would the traveler be so small as to be able to avoid all the debris? If the traveler were 6 feet tall on earth, then the traveler would be 3.2 inches in length when traveling (at 99.9% of the speed of light, remember), so maybe the traveler would be small enough to avoid all the debris, depending on the risk factor - and I'd set 'a very safe length' at .000001 inches (and this when the average distance between 'space debris' is around a million miles or so).

At **99.99% of the speed of light**, the traveler would be 7 tons and only 1 inch in length, which is better, but not ideal; there would have to be a lot more 9's in the speed decimal place for me (closer to the speed of light);

so let's put six 9's in the decimal place - **99.999999% of the speed of light**, bringing the travel speed up to one millionth away from 100% (the traveler cannot travel AT the speed of light - at that speed the traveler would have a length of zero and a mass of infinity - no can do). So at 99.999999% of the speed of light, the traveler's mass would be 170,000 tons with a length of only one hundredth of an inch - not too shabby - the traveler would bore right through any space debris with hardly a scratch...
and, if you're curious, the traveler would only have to travel 7 days (14 years for the waiting person), which reduces the opportunity for such risks.

Acceleration Forces
There are a lot of planetoids the traveler could visit out in the Oort Cloud, if the traveler could only slow down

and speed up in a reasonable time, which is unlikely, given accelerations forces on the traveler's frail biological body (we'll work it out)... so maybe a very high-speed shutter camera would be more practical - capturing a few pictures in passing, when set on proximity auto-snap and correcting for length dilation...

Let's say the traveler's frail biological body can take only 10 g-forces. What kind of acceleration is that - meaning how long would it take to reach 99.9% of the speed of light? At 10 g-forces, it would take a mere 35 days, so acceleration is not much of a factor - or is it? Stopping for a planetoid visit requires 35 days for deceleration, and 35 more for acceleration again, adding 70 days to a 9 month trip. This could screw everything up - why, the waiting person would age another 5+ years! So a shorter trip (faster speed) would entail more hazards, and small glitches in one's plan would have a greater effect on the desired outcome.

You can see how limited acceleration and deceleration would add immensely to a 7-day trip (and complicated time dilation calculations) - it would add nearly 70 days (35 for acceleration and 35 for deceleration (minus dilation during acceleration and deceleration periods), so a minimum trip would be roughly 60-70 days, which means the traveler would have to travel at a slower near light-speed. If a 60 day trip were planned, for example, then the near light speed would be set at only **99.993% of the speed of light** - the traveler might have to turn on the emergency flashers, the speed would be so slow....

Journal 145: Pondering Alternate Cosmological Constructs

Let's say the Big Bang happened, and we, as life, are on a wild ride down the energy gradient as the resultant universe's energy dissipates and disperses - for it is said that phenomena like the 'molecular storm' (which powers the molecular machines that do the work to keep us alive) is merely the remaining energy - so every time you take a breath, you are 'using up' part of that lingering energy, until it is all used up and dispersed.

So then what? Well, it may be that it just so happens that we exist because of the varying independent forces at work in our physical world - and 'independent' is critical. Let's say that all the matter in the universe, every last subatomic particle, finally 'evaporates' in a slow burn of atomic radiation. Why is that not the state of the universe right now? How are we here? It could be that gravity is the working force that 'recharges' a dissipated universe - pulling matter together until there is enough for the next Big Bang (the 'recharging') - via a singularity that goes around sweeping up the debris of former Big Bangs, until the singularity becomes so large that it becomes unstable and [insert an silent expansion here] the forces holding it together is overcome by the forces in favor of expansion.

Now weigh this against the next thought I had: that matter accumulating together via gravity may be simply 'falling into smallness' - in that the general direction of matter is to 'become small' - as if everything in this universe is falling into a smaller one (or multi-mega-illions of smaller ones - as if we were 'too large'.

Now if this process is a general law of nature, then we can expect 'universes' to become smaller and smaller. The problem may be that the process is not perfect - perhaps there is a rebound effect, or a saturation effect, that pushes back within a singularity which is that root cause of it suddenly expanding. So perhaps 'some' of the total matter falls into the next smaller plane, while some is ejected outward back into the universe it came from, to slowly dissipate in energy again, and be swept-up again by a passing singularity, and either to 'fall into smallness' (become part of a smaller universe) or be ejected outward again.

Journal 144: My Favorite Shapes and Curves

These particular shapes and curves, and let's include diagrams, are among my favorite because they are so useful - for they can depict many things that I am endeavoring to communicate while developing a post-religion, new 'world-worthy' philosophy.

The Triangle

Example Application: Reduction of Needless Conflict
This one is so new, so simple, and so important (in reducing conflict) that it needs a catchy name (so it can be instantly remembered by those contemplating intellectual contention with or violence toward competing theorists) -

So let's call this the "INVERSE DATA LAW". Got that? The Inverse Data Law.

Inverse...
Data...
Law.

Which states (in an easy-to-visualize pictorial diagram) (using one of my favorite shapes) that, *"The less data there is, the more there will be (and the more we*

need) diverse hypotheses" - (but let's call the hypotheses 'theories' since hypotheses are popularly known as 'theories', even though they are technically different) (hypotheses being initial explanations in need of further testing, while theories are tested and accepted models (sometimes consisting of many hypotheses) that 'work' (are 'adequate' as I like to say philosophically), though they are not yet completely understood - the Quantum Theory being a good example).

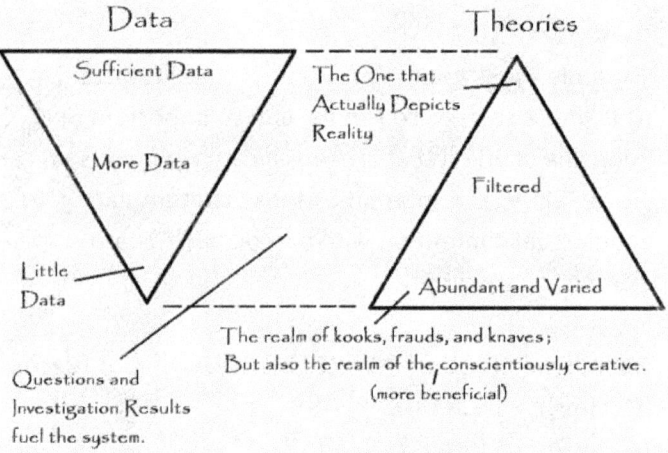

So these triangles depict the relationship between data and the number of theories that attempt to explain it. As you can see, the less data there is, the more theories there can be. As more questions are asked and the investigative results obtained and input into this system, erroneous theories are exposed and discarded. An adequate amount of data is defined as that which begets the theory that actually (or 'adequately') explains reality.

Why is this diagram so important? Consider scientists, how they needlessly and viciously attack each other and each other's theories while the theories are still in the developmental stage - i.e. when there is still little data to go on. If the scientists knew of my simple discovery depicted here by these amazing shapes (triangles), then they would realize that not only are many theories expected at this stage, but desired - kooks, frauds, creeps, comics, and knaves aside (the usual detritus found floating in a theory-rich environment) - but at this stage, even they contribute positively, if only in a negative-sum way (that is, placing more urgency in obtaining more data).
So, 'best case', this diagram, based on the simple triangle, will eradicate needless conflict in the scientific community, and will promote cognitive creativity, which is essential and beneficial during the early stages of theorizing.

Example Application: Alternate explanations of Egyptian archeological finds,
https://www.youtube.com/watch?v=vaXLOIGdUOc
Here we have little data to go on, and a creative theory arising.

As data accumulates, alternate theories dwindle. This video indicates one of three states - 1. That there is still little data; 2. There is an ignorance of existing data; 3. Both. The video statement "we don't have the technology to build the pyramids now" answers that question - 2 (ignorance) and 3 (little data), but preponderantly 2 - there is an ignorance of existing data (for today's technology can easily build such pyramids, and with far less manpower, and in far less time - so there is a general disconnect with reality in the theorist, which, most likely, carries over into the theory).

This video also illustrates the danger of getting caught up in creative theories - you then dismiss the existing body of knowledge (existing data) out of hand, and you fail to educate yourself, getting instead caught up in theories whose only merits are that they are speculatively exciting and intriguing (or worse, simply attractive and fashionable - so much so that people 'wish' they were true - my Fuzzy Blue Neutrons are a case in point (I am still wishing hard)).

On the other hand, the best services that alternate theories can perform is 1. To give science new questions to answer (for science is all about answering questions, after all), and thus opening up new avenues of worthwhile investigation; and 2. Added tools of perception - meaning tools (varying perspectives) with which to peer into the unknown with.

Are the theories here worth investigating? Yes, only because all theories are - it is just that an errant theory only takes a minute of investigation to dismiss in light of existing data. A weak theory on the other hand (as opposed to simply 'errant') can be filed away in the near endless rows of cabinets stuffed with other speculative possibilities, pointless to draw conclusions when there is still not enough data to go on, but retained as still possible (and in the meantime useful as the stuff of good science fiction).

Example Application: **Reduction of Suicide**

Preventing Suicide ~
Cause #23: Being Born
Into a Nice Environment

The Problem With Being Born
Into a Nice Environment

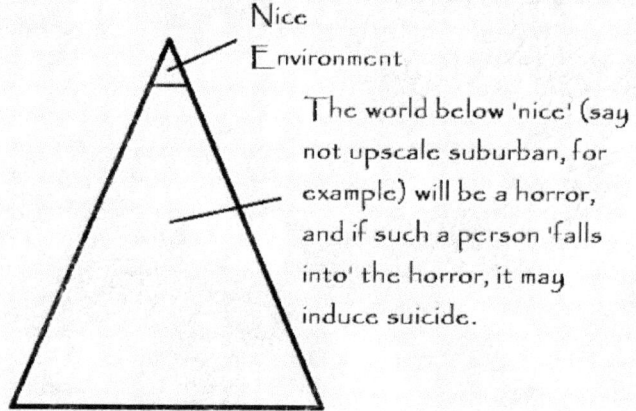

Nice
Environment

The world below 'nice' (say
not upscale suburban, for
example) will be a horror,
and if such a person 'falls
into' the horror, it may
induce suicide.

There are three solutions:
1. learning how to deal with the
horror via first-hand 'exposure'.
2. being on business - which reduces
idleness, which reduces 'trouble'
3. Actually caring about those who
were born in the horror.

Here we use the triangle to help communicate a
different concept - how some people are born into

nice circumstances, and to whom most of the world is then a horror - which, if exposed to, may induce suicide in the 'nice' person, which, we can all agree (if we all acknowledge my new 'world-worthy philosophy', which holds that the more minds that can be potentially applied to a problem, the better), would be a loss.

Conversely, a person born at the bottom - meaning into horror, sees only 'up' - where most of the world 'out there' (beyond their immediate horrible circumstances) is like heaven, and to these people most of the world holds potential happiness.
You can deduce that this can all be reduced to 'perspective'. If you lack perspective, you may not be able to deal with that which is beyond you, and you are more susceptible to suicide. Sad, since it was merely perspective that was lacking.

The Oscillation

Example Applications:

Social Change and Global Climate

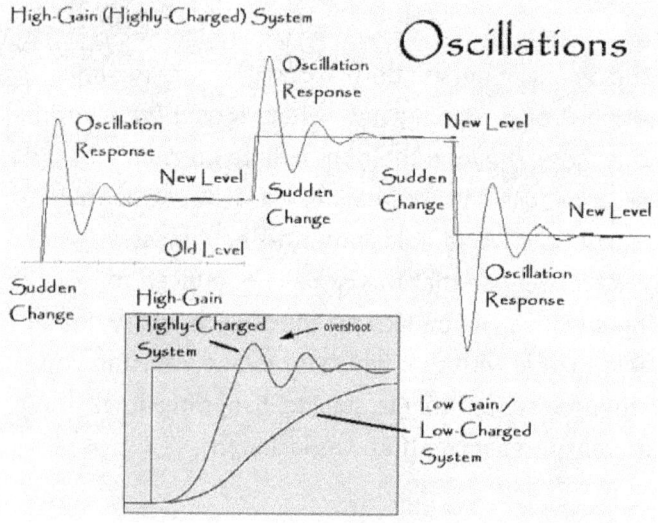

As for the various types of oscillations - random, continuous, and damped, the overshooting then damped oscillation is most useful for my application, for this can represent many systems where a sudden change can occur, such as social systems or earth's climate. Let's look at each application.

Social Systems

In a social system, voting can be represented by this oscillation, as it swings back and forth from Left to

Right. You can see that when there is a sudden change, and if the system is highly-charged, there may be 'overshoot', where the new desired point will be passed (extremism), and will wildly swing between extremes, but then eventually settle out at the new desired point, though unfortunately after many needless deaths. Another sudden upset, another round of oscillations, another round of needless deaths.

Global Climate
The same can be said for the earth's climate. If an upset occurs, sudden or otherwise, based on how charged the climate system are, the climate may, in response, swing wildly before settling at the new equilibrium. Another upset, another potential round of wild swings and gradual settling.
Conversely, in a low-charged system, the change toward the new equilibrium will be slow; in fact the system may never reach the new equilibrium point before the next upset occurs.

The Bell Curve

Example Application: **Allocation of Resources**

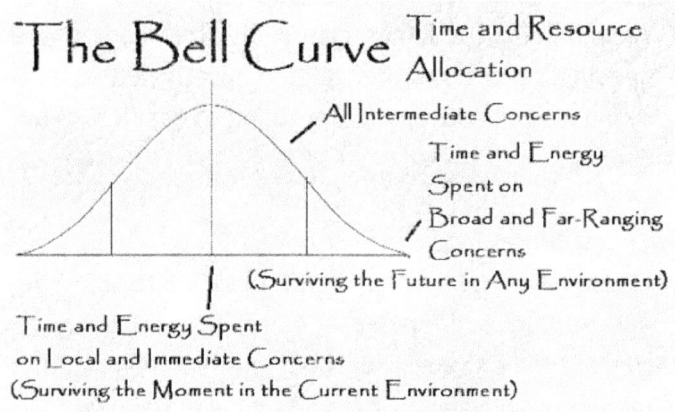

The Bell Curve — Time and Resource Allocation

All Intermediate Concerns

Time and Energy Spent on Broad and Far-Ranging Concerns
(Surviving the Future in Any Environment)

Time and Energy Spent on Local and Immediate Concerns
(Surviving the Moment in the Current Environment)

In my Species Brain Age hypothesis, I've identified 'primitive activity' (and mindsets) as those that are merely concerned with the 'local' and the 'immediate', as opposed to the more advanced 'broad' and 'far-ranging' (which includes the future). The problem is, you cannot be 100% advanced - for local and immediate concerns will always exist, though the will become less and less of an issue.

Also, a species that is well into its Brain Age will become proactive in discovering, and then dealing with, threats to life in a universe, in order to proactively secure life (and the higher priority, higher consciousness - that which is proactive) in a harsh universe.

The question then becomes how much time and energy should the species spend on discovering ever-broader, ever-far-ranging threats to life - and how much time and energy should be spent on the local and immediate, and the answer can be depicted by the Bell Curve.

In normal circumstances we can expect to spend more time and energy on local/immediate concerns, which is depicted by the center (most voluminous) portion of the curve. The allocation of resources can be guided by this applied curve, such as government and corporate spending, or a person's individual attention over time.

The Scatter Diagram

Example Application:
Diversity Monitoring

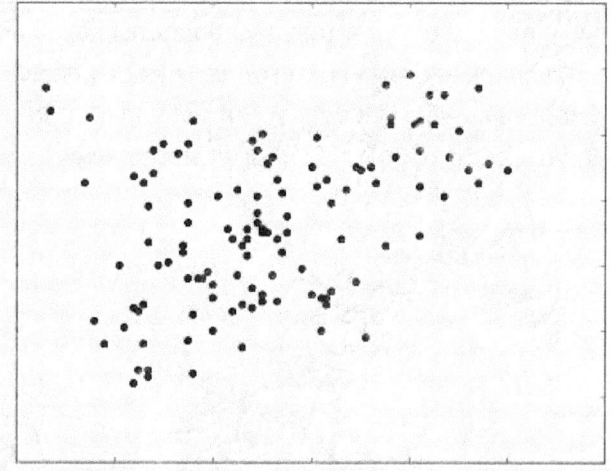

Scatter Diagram ~
Diversity of Life

The broader and more evenly diversity is scattered, the higher the probability of the survival of life.

As applied to maintaining/creating adequate diversity for the continued survival of life, the scatter diagram is the diagram of choice.

Delving deeper into this particular application (since it is so important), the question becomes, "What measures are to be applied to 'measure', and thus place, diversity on this diagram?"

One can readily imagine that there are many taxonometric features that can be used, everything from genomics to environmental niche adaptation. Philosophically speaking, it can be seen from the diagram that if one physical form spanned the entire diagram, then it would cover all possible diversity in itself, and such a form would be the 'Ultimate Form' in terms of guaranteed survival (if indeed a physical form is the answer).

As it stands today, the best odds for the survival of life and of higher consciousness (proactive) is diversity - both physically and socially, since an ever-higher consciousness is ever-more dependent on ever-more complex social factors.

Tolerance

What this means is you will develop 'tolerance' toward entities that are different from you through the understanding of, and recognizing the value in, such new insights embodied the prospective 'new world-worthy' philosophy (which is still in development - so roll your sleeves up, proactive higher consciousness that you can be (your choice).

The Pie Chart

Example Application:
Society Analysis

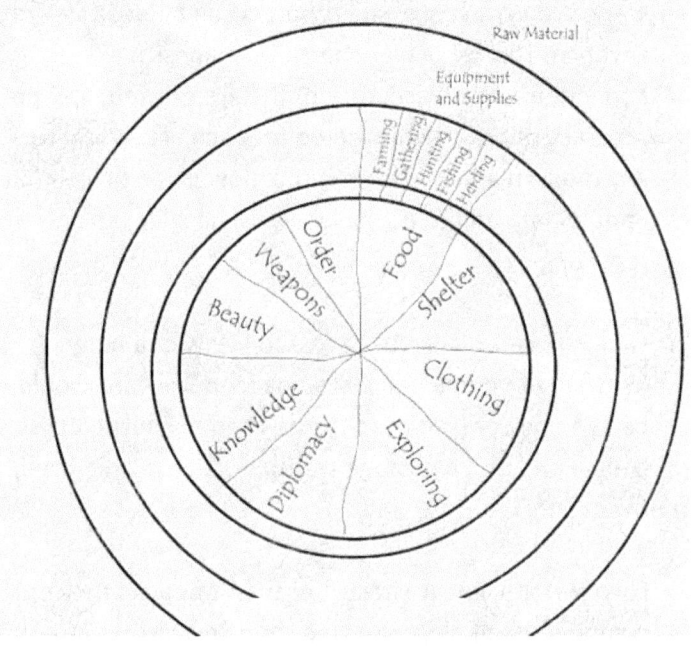

Concentric Pie Chart - Society - Layers

This is not your ordinary pie chart - this one needs a new name, such as "Layered" or "Concentric-Relational" - serving deeper relational analysis.

I would use this approach in conjunction with economic and census data to identify voids and needs that are under or over represented, from which I would then inform the public as to their economic peril (if involved in an over-represented area) or of opportunities (regarding under-represented areas).

For example, let's say that there will be an impending shortage of farming equipment and supplies. If identified in advance, the problem can be dealt with in advance. This particular chart's usefulness is dependent on the accuracy of the areas listed, and the accuracy of the data collected for each area. As with any modern analysis, computing power can be used to update data and to do the analysis, flagging, and reporting.

I like it because I like to have a full view of a body of knowledge on one quick-scan page. A 3x5 card would be even better (such as the one I had my entire physics course on), and a postage stamp is ideal - all digitized, of course.

I like the concentric circles because it depicts the core components of a system, and then, in successive layers, all that contribute to their functioning. Here, for example, you have 'Food' as a core component of a society, then various methods one layer up, 'Farming', for example; then the various supporting activities for each method, such as manufacturing equipment and supplies, and to support that, the discovery, gathering, and processing of raw materials.

How do societies currently function? Two ways have been tried (one failing and one succeeding): 1. Centralized control - which failed (the various communist experiments), and Capitalism, which, valuing individual initiative and enterprise, succeeded, though imperfect - people learn of over-representation the hard way - when they get to market and they find it saturated, and people learn of (and then service) opportunities only when there arises a clamoring for such services and support. A government serving the people can use the chart above as a central monitoring service (or better, a private company - reducing the size of government), with the goal of reducing market saturation's by identifying over-production in advance (such as a lot of one seed selling and little of another), and precluding under-production by identifying opportunities in advance (based on projected production/usage data).

Journal 137: The Future of Paleontology - Inference with Ever-More Microscopic Evidence

Bones! Complete extinct organisms impressed in stone and frozen in tundra! Trilobites! Ancient fish skeletons! Precambrian Fossils! What do all of these have in common? A few things: 1. they are what current paleontologists are looking for; 2. they are all BIG.

While it is nice to find complete skeletons of past creatures, in many cases it is not possible - for example those that lived in forests, where quick decomposition rules. So what does a paleontologist do? The answer: Search for ever-smaller evidence - and the smallest would be on a molecular/cellular level - the goal being finding the genes of a past creature.

So, in theory, a paleontologist could dig down forty feet into the layers of biological debris left behind by living entities over the last three-and-a-half billion years, and analyze what is there - most likely nothing 'big' will be unearthed, but there may be a rich record on the molecular level, even genetic level.

Therefore a future paleontologist's primary tool may become the electron microscope.

Then again, maybe only for a limited range of history - for it is now said that the half-life of DNA is only 512 years, so the oldest DNA that would be readable would be only 1.5 million years old - far short of the dinosaurs (65 million years ago), let alone earlier creatures. So rock impressions and fossilized remains are still an important pursuit, but like technology, everything is shrinking...

Journal 135: The Bell-Curve Model for Mental Focus and Government Funding

We can continue to let politicians splay our money in haphazard directions, or toward the loudest lobby groups and special interests, or we can let them continue to bribe us with our own money...

Or we can give them guidance - which is what I will endeavor to do right here.

Consider this Bell Curve:

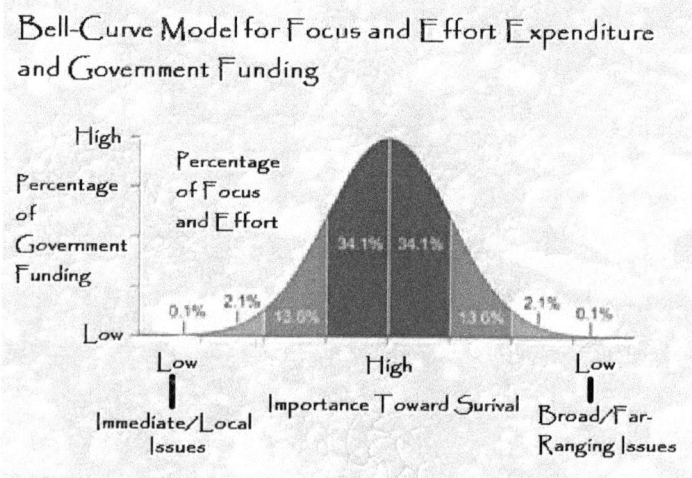

Bell-Curve Model for Focus and Effort Expenditure and Government Funding

You can see that the horizontal axis spans issues that are important to survival, spanning issues from the most immediate and local to the broadest and distant, where the issues with the most impact fall in the middle. I should give you examples of each.

Immediate/Local Issues
These issues are either self-centered or dealing with the moment, such as how you comb your hair today or what nail polish you will wear. I don't mean to trivialize such things, but there are not important toward the survival (and perpetuation) of life (though females may argue - they need to be attractive and in style). Entertainment would be placed in the local/immediate area, though as an art it could span all the categories, from local/immediate to broad/far-ranging.

Broad/Far-Ranging Issues
These generally exist at the frontiers of science - how the universe came into being, how it works, where it is going in one-hundred million years. Such knowledge will spur us to deal with long-range survival plans.

All the Other Issues In-between
These issues are issues that you will face in your lifetime - family issues, society issues, political issues, environment issues, and they are issues that you may have an effect on in your lifetime, and where you would still be alive to see the results.

For example things like national defense suck up a lot of focus and energy that could be better spent on broader, more far-ranging issues - but we are still crawling out of the Stone Age on that social frontier.

Journal 134: Earth Climate Change - Feedback Loops and Oscillations - A Total Predictive Model

My title is facetious - a total predictive model does not exist yet (nor am I about to spend the rest of my life pursuing one) - the sum-total of earth's loops and oscillations are as-yet unpredictable. When we can identify all the inputs to "the system" (earth's overall climate, which affects local weather) and their characteristics (direction, period, and intensity), then we can build an adequate model; until then, we deem such system 'pure chaos'.

Therefore, if you read someone who makes the claim (or any claim) that our greenhouse gas emissions are going to be detrimental, you know you are dealing with bunk, with twaddle, with charlatans, with spotlight-pirouetting frauds, with mountaintop prophets severely detached from reality.

To cover the details, what makes such predictions unfounded (and currently impossible) are the many diverse and local climate feedback loops and oscillations that exist on earth (and those which exist in earth's vicinity - meaning the sun), and which all interact, continuously, in real-time, making any prediction (especially with the nascent information and technologies that we have at present) disingenuous at best. One would suspect such predictions were derived from divine murmurings, or were just plain contrived (to be redundant).

Earth's Climate Oscillations and Feedback Loops

On earth, each oscillation and feedback loop exists at a different location (for example those emanating from rainforests, or the Arctic, or the Antarctic, or a desert, or mid-ocean), and all loops have different gain effects (intensities), and all have different time durations, making the sum of their interactions impossible to predict (as our understanding now stands). For example, Fourier analysis of geological data has not shown any identifiable patterns over time - bringing into question natural climate-affecting oscillations themselves.

Considering the infinite number of ways all oscillations and feedback loops might interact at any given point time (all classified as 'inputs' to 'the system under study'), and how the earth will react over time, and taking non-periodic events into consideration (such as present human greenhouse emissions or a major asteroid hit) (and we still cannot see them coming), it is no wonder we have not mastered global climate understanding (let alone effecting control or modification).

Various Possible Earth Reactions to a Step Change in Temperature

Feedback loops have been long used in industry, yet I do not see any climatologist referring to the standard knowledge gained there*. If any have, wonderful, but I've found none, so I will use that knowledge here and apply it to climate change.

In industry, the method of 'controlling' the response of a system is GAIN, and MODES OF CONTROL (usually more than one working together, where each introduces gain differently (such as proportional, integral, and derivative modes).

As for earth's climate, the overall gain would comprise all the inputs from the rates of change, directions of change, and intensities of change of all the local, independent loops and oscillations around the globe.

A graph is worth a thousand words (and is a superior communicate method than relying on interpretive mental imagery based on words):

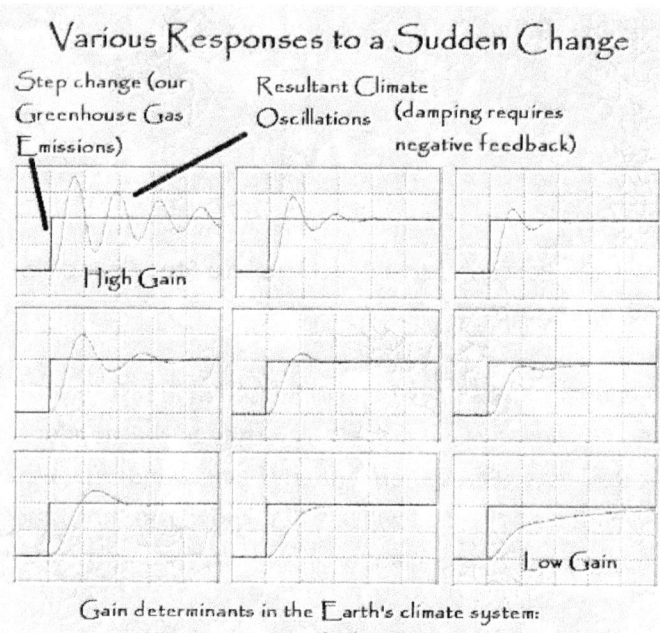

Various Responses to a Sudden Change

Step change (our Greenhouse Gas Emissions)

Resultant Climate Oscillations (damping requires negative feedback)

High Gain

Low Gain

Gain determinants in the Earth's climate system:
the contribution of a feedback or oscillation,
determined by direction, rate of change, and intensity.

The graphs depict responses to a "step change" (the right-angled line), which in our case represents present human greenhouse gas emissions, while the curved line represents the total system response (global climate). Note that in industry, the step change is called a 'set-point change' - and it is a 'desired' point (in climate, it would be our desired point). The trick is getting the system to the new point (and maintaining it at a desired point) - without the plant exploding, or in our case, without causing the extinction of all life on the planet. Such control relies on gain, modes of control, and feedback.

What doomsayers choose to ignore (or simple haven't realized yet) is the most important point - 'WHERE' in time - meaning where in the waveforms of earth's climate change - are our greenhouse gases are introduced. In the chart below, you can see that Line B represents negative feedback, contributing to system stability. If our actions have positive feedback, however, such as at Line A, then we could drive the earth's climate system into extreme weather conditions.

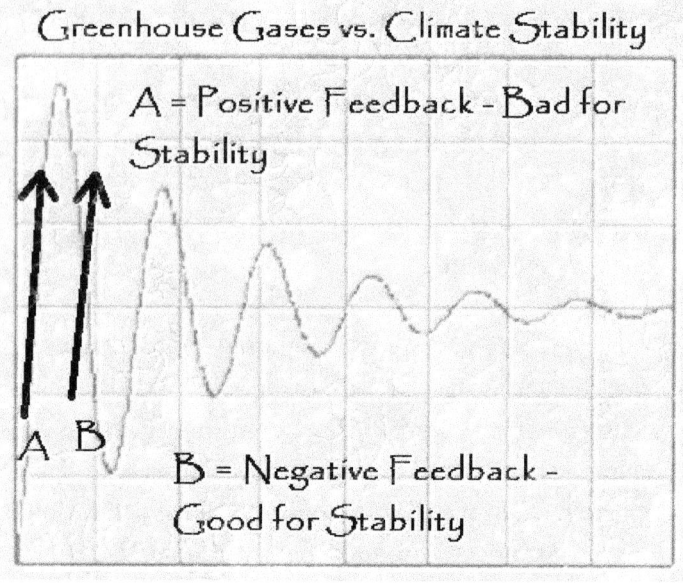

Greenhouse Gases vs. Climate Stability

A = Positive Feedback - Bad for Stability

B = Negative Feedback - Good for Stability

In summary, releasing greenhouse gases at point A will contribute to earth's climate instability, and will lead to extreme weather cycles, while gases released at line B will oppose whole-system rate of change, contributing to stability. This is the key consideration for future climate control (as opposed to 'weather' control, which is non-global and localized), once we have an adequate knowledge of all the independent systems involved.

There is one response not shown above - gradual runaway:

Runaway Response

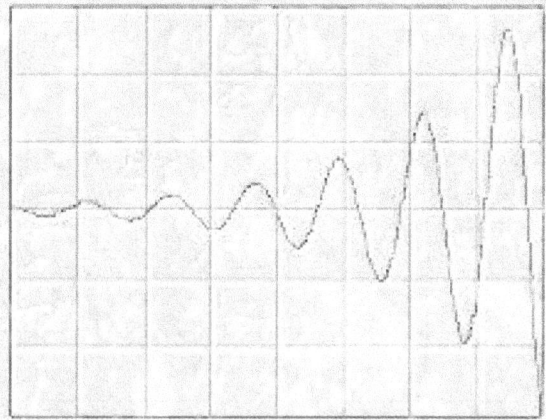

Caused by gain and positive feedbacks being slightly too high for adequate control.

To get this system back in control you must introduce negative feedback or turn down the gains, which will serve to dampen oscillations.

Here is an interesting out-of-control state - where the high-low extremes extend beyond a system's extremes:

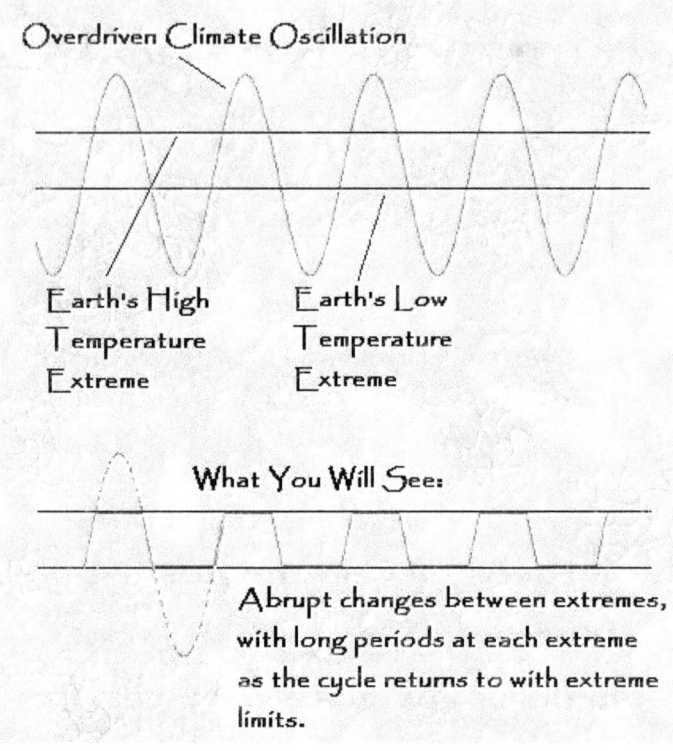

Overdriven Climate Oscillation

Earth's High
Temperature
Extreme

Earth's Low
Temperature
Extreme

What You Will See:

Abrupt changes between extremes,
with long periods at each extreme
as the cycle returns to with extreme
limits.

Once again, how long the system will stay over-driven
and oscillating will depend on negative feedback.

Just a few interesting notes concerning an over-driven
signal - in computers, square waves (used in bit/byte
communication) are created this way, and in audio, it
causes audio distortion (volume up too high).

Summary

I'd like to sum it all up with one pictographic depiction of the complexities involved - I have in mind a photo of the earth and sun from space, and the locations of the various earth and sun feedback loops and oscillations.

I won't name them in the picture (it would become cluttered), but various generalized feedback loop and natural oscillation examples are volcanic eruptions, basalt flows, hurricanes, organic matter decomposition, peat decomposition, permafrost thawing, rainforest drying, forest fires, natural methane releases, desertification, CO_2 in oceans, cloud feedback, ice and glaciers, water vapor, chemical weathering, and blackbody radiation to rattle-off a few (and in different locations they have specific local names, such as the Greenland Ice-Albedo Effect, El Nino, Thermohaline Circulation, the Beaufort Gyre, the Hudson Bay Permafrost Thaw, and Boreal Forest Carbon Cycles (to give you a few names to drop at the next gathering of prospective climate change enthusiasts).

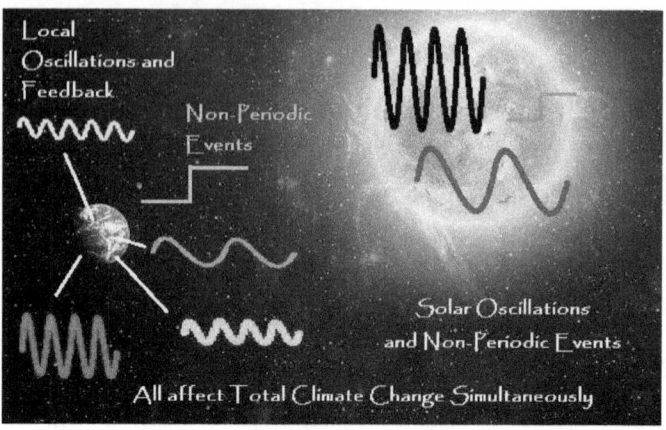

Local Oscillations and Feedback

Non-Periodic Events

Solar Oscillations and Non-Periodic Events

All affect Total Climate Change Simultaneously

You can see how we do not as yet have a grasp on the entire system - we haven't identified all the components (inputs to 'The System') yet, not to mention being able to handle the effects of major non-periodic events, which would play havoc with the entire system.

So people predicting dire consequences have their mortarboards up their asses (my term, coming from industry) (and if indeed they have one - for in today's media-sensation times, this is not a requirement).

In short, the current human greenhouse gas event is but one factor in many, and, depending on what point in the system it is occurring, and its intensity, and its relations to other components (feedback loops and oscillations) and their parameters, it may have either stabilizing or destabilizing consequences.

What Do We Do?

Do we 'play it safe' and cease our creation of greenhouse gases, or do we try to understand the system as a whole, where we can come to modify, if not control it - or at least know what the predicted effects will be, based on the entire model? My preference is to move toward understanding and control - leaving the doomsayers to their own nefarious goals.

Journal 133: Preparing for Interstellar Travel - Reducing Our Biological Size

One of the most ludicrous scenarios I've come across from space science fiction (as it grapples with the details of future space existence) was the idea of having a pasture, complete with pasture-grazing animals, in a habitat space vehicle around three hundred miles square (around ten by thirty miles).

If that does not present a ludicrous mental image to you, then let me explain:

The size of the space habitat is not the issue - the issue is us (and animals) - we are, and animals are, huge and inefficient - requiring extraordinary amounts of fuel and creating huge amounts of wastes (compared to microbes) - and thus we (and animals) are simply not, at present, suited for space travel/existence in a self-contained enclosure - we are too inefficient.

You would immediately counter with 'recycling', which would solve the problem - but it may not be chemically possible. If it was, and it could all be figured out, great - size does have its advantages (but it also has its disadvantages - just consider our susceptibility to viral infections).

The path that I was thinking along, as can be seen by the title, is reducing our biological size solely in order to minimize our energy requirements and waste production.

Thinking about this, I've come upon a few disadvantages, of which I will cover here:

Problem - Alien Competition
If we encounter alien competitors that are 'bigger' than us, they may have their bullying way.

Solution - Swarming
We could counter by 'swarming' - like a hive of bees, or by infection, as with viruses - in effect, acting as one larger unit.

Solution - Technology
As we progress down in biological size and up in biological efficiency, our technology would become smaller, making it less competitive. But as we progress smaller, so too should our mastery over ever-increasing sizes of technology. In this scenario, even though we were much smaller than we are now, the technology at our command would be much larger. True, if we stayed large, our technology would be even

larger, making it more competitive on a galactic life stage.

Consider the size of a commercial airline pilot with the size of a jumbo jet - you have a (relatively) puny bio-entity in complete control over a huge system of technology, and this size ratio is ever-increasing..

Problem - Brain Size

If we reduce our size, this includes reducing our brain size (unless we go for brain-only disproportion), so the question here is, "How small can our brains be while not losing any cognitive power?" As it is, there seems to be a lot of room for improvement:

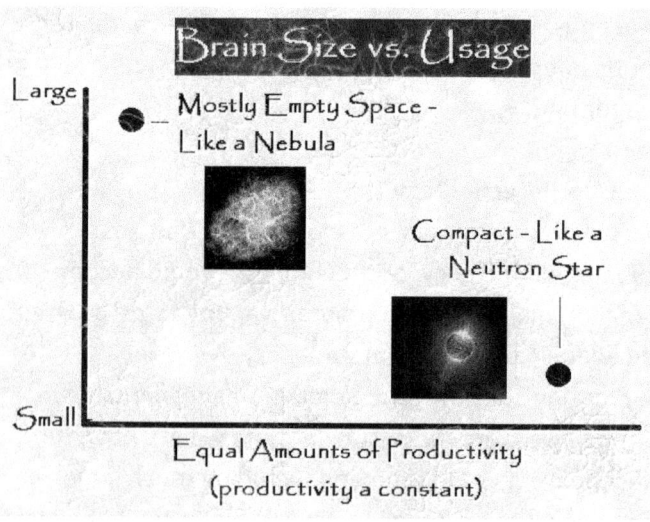

Brain Size vs. Usage

Large

Mostly Empty Space - Like a Nebula

Compact - Like a Neutron Star

Small

Equal Amounts of Productivity
(productivity a constant)

Solution - Sacrificing Biological Senses
It appears that a large portion of our brain is tasked toward sensory reception - sight, sound, touch, smell, taste. Since we have already enhanced our more critical senses with technology (sight and hearing in particular so far), it would seem that we could sacrifice the related biological senses, thereby rendering large portions of the brain expendable.

Solution - Hard-Brains / Bio-Brains
We could transfer our cognitive power to hardware - and if microelectronic in size, the energy requirements would be low. There is also work being done on creating computative power out of biomass, which would lead to 'bio-brains' - brains custom-designed for our needs.

Solution - Social Coordination
We could just skip all of that, and sacrifice individuality, where we would function as a group - turning our meager individual cognitive powers into specializations - sort of like what cells have done with our present bodies - individually, a cell is not a formidable cognitive force, but working together, they have created us (though we can argue the claim of 'formidable cognitive force').

Problem - Shorter Life Span

As a general rule, the smaller the being, the shorter the lifespan. There is another general rule - that all beings with a beating heart have roughly the same number of heartbeats per life - but since smaller beings have faster-beating hearts, their lifespans go by faster.

Solution - None

At least that I can see - for if you consider life-lengthening research, sure, it can be applied to your smaller self, but it also can be applied to your larger self - where you would live an even longer life, and theoretically be able to accomplish more... maybe...

Advantage - Speed.

Just like heartbeats, it may be that we will be quicker at thinking and accomplishing things when smaller, so in a shorter lifespan, we would accomplish more in a given amount of time, and in the end equaling what our slower selves would accomplished in their longer lifespans. So if the measure of life is what is accomplished in it, your small and large selves would end up accomplishing an equal amount. Such speed also translates into strength, as in competition with aliens. The best solution for a species in general is (drum roll) once again 'diversity' - having a variety of small beings and large beings, taking advantage of the benefits of both, and learning to deal with the disadvantages of both.

Problem: Predators
If we are smaller, we would be in hazard of becoming prey to the predators on earth, even small insects.

Solution: Take Progress In Small Steps
This will give us time to formulate adequate measures for our own survival - which is critical - for higher consciousness gets priority in the grand scheme of life - since we do not have eternity to develop (the sun, for example, as a limited 'friendly' duration).

Journal 132: A New Way to Mathematically Describe Any Object

Introduction

Fred Hoyle compared the probability of life originating from non-living matter to that of a hurricane blowing through a junkyard and assembling a Boeing 747 jumbo jet.

This is a bad analogy, and a false premise (given laboratory experiments in microbiology), but that is not why I mention it here - what it indicates is that we can (applying reductionism) 'reduce' the description of any object into the sum of all the forces that were required to create it (and in the proper sequence, if sequence is a critical factor).

So when we mathematically describe any object, we can describe not only from a present-state matter and energy point of view, but from the perspective of the forces required to create it (and maintain it - though that gets into 'present-state').

Analyzing the Hurricane

Let me analyze the hurricane and jumbo jet to illustrate. A hurricane offers a lot of force, but it lacks many of the forces required to assemble a jumbo jet - rendering the probability nil.

For example it lacks 'multi-revolutionary torque' (which is required to screw-in a screw) and force transduction abilities - that of converting one force into another in order to accomplish a task: For example transducing electricity or a chemical into heat in order to weld. The hurricane has forces, but not all the required forces.

This is not to say the hurricane could not obtain such forces through secondary means, such as spinning bit of flying metal that haphazardly and by miraculous chance screws in a screw, or causing an electrical fire hot enough to weld two passing pieces of metal together in just the right places, and in sequence in relation to all the other tasks involved.

In probability, it is easy to get very high numbers against something occurring, astronomical, and indeed, hyper-astronomical (meaning more than all the atoms in the universe) (at least the observable universe), which, in math terms, is around ten to the power of eighty or so (and remember, it takes ten ten-to-the-power-of-seventy-nine's to make one ten-to-the-power-of-eighty - just thought I'd mention that -

so we are talking about a lot of atoms, and a very slim probability if hyper-astronomical).

So you can see the probability against a hurricane assembling a jumbo jet rising 'astronomically' when we get into secondary forces (those beyond the hurricane's innate forces - which are more or less linear in direction and constant in force over the given short time-span of a typical 'task').

Summary
You are a result of many forces, and you can be 'mathematically described' by those forces, the aspects of which not only include quantity and variety, but also time and location (relative to something else, of course). Worthy Pursuit - inventing goggles that can 'see' this.

Journal 131: On Intelligence and Evolution

Nutrients

Evolution is habitat-driven, and the primary driving force is obtaining which nutrients (to keep the probability of life-extending events happening within sufficiently high - see my Chaos Theory of Life. Evolution would not be needed if nutrients were plentiful - life could exist perpetually in a microbial state - there would be no advantage to growing larger or stronger or faster (in fact they would become nuisances, if not handicaps - requiring more nutrients). There would also be no advantage in acquiring new senses such as sight, smell, hearing, taste, or touch, or no advantage in acquiring higher cognitive abilities such as memory or cause-effect reasoning, or faster response times via improved nervous systems.

Wastes

If the habitat simply washes away biological waste, and if there were no predators, then there would be no need for biological evolution. If the wastes lingers, however, and begin to kill its creators, then something has to change, or that creator will go extinct - and 'evolution' will 'select' those organisms that (by pure chance, see my Chaos Theory for Life) happen to find a way to escape their own wastes (and by the same reasoning, predators).

Mobility and Physique

I've covered the advantage of acquiring mobility and greater physique in the evolved 'hunt' for nutrients (as opposed to being stationary and waiting for nutrients to come along), but it also allows one to 'get away' from wastes (at least upstream) and predators. The problem here is that acquiring greater mobility and physique also includes creating greater wastes (more 'fuel burned'), which may counter. if not negate, any advantages gained in physique; and mobility may also attract predators - many of which depend on sensing movement and vibration (in addition to chemical stimuli).

Senses and Cognition

The same goes for senses and cognition. The advantages of acquiring new senses and cognitive abilities in enhancing the ability to obtain nutrients are obvious, as well as their playing a part in getting away from biological wastes and predators. Scenario: Imagine a stationary organism with stationary nutrients 'just out or reach'. Evolution would favor that organism that (by pure chance, see my Chaos Theory for Life), developed a way to 'go to' the nutrients rather than waiting (in this case in futility, and fatally).

Natural Evolution vs. Intelligent Evolution

Natural (biological) evolution leaves all this to chance, while intelligent evolution identifies the important factors to life (such as securing nutrients and dealing with wastes), and takes PROACTIVE measures in finding solutions. If you relied on natural evolution, you may have to wait another hundred-million years to solve a problem. If you use intelligent evolution, it may take only decades. I call this transition a specie's Brain Age - an term that encompasses the use of reason and intelligent evolution (which includes social structures and biology-enhancing technologies).

Terminology

Rather than say 'evolution', I like to say 'progress' - because the term 'evolution' is still controversial (religious types like to attack it, or ignore anything that mentions it), while 'progress' (particularly technological) can be seen first-hand and in action by everyone (and is less controversial) - and seen within one's lifetime (while evolution remains a theory - derived from reason, and occurs over hundreds of thousands (if not millions) of years); and the only people who see evolution first hand are breeders and biologists - not a large segment of the population. Social and intellectual 'progress' can be argued against, since 'change' can be trivial (fashion) or detrimental (decadence), though I've covered them in detail elsewhere (in my Brain Age, New World Philosophy, and Artificial Intelligence journals.

Summary

'Progress' (and let me (in a whisper) 'evolution' here) can be measured by the breadth and depth of threats (and benefits) to life that we've discovered and have dealt with (if not put to use), either passively (letting biological evolution take its course) or proactively (using our brains to seek and solve). This means that there is a concrete measure for progress (when compared to extraterrestrial beings, for example - for otherwise there would be no need to 'measure'). In this line of reasoning 'progress' is not debatable (again, as are many forms of pseudo social 'progress' such as style or fashion - which may work today but not tomorrow (and by 'work' I mean enhancing your perceived status or attractiveness (the only reason style and fashion exist - just to move you up the perceived intelligence curve temporarily).

Intelligence, Evolution, and Our Predator/Prey System

Here is my pet scenario on how our predator/prey system came into being:

One future predator to another: "Hey - all those little bastards ate everything! Now what do WE eat?"

Journal 128: On Relative Alien Intelligence

Consider the following chart:

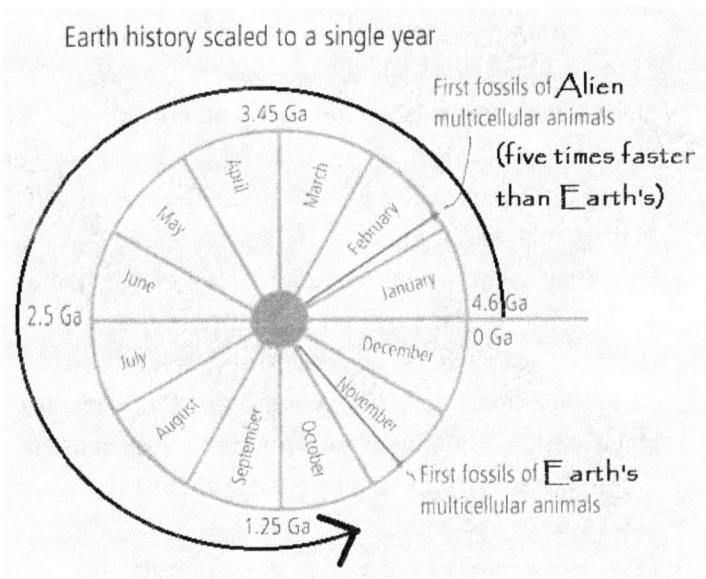

Earth history scaled to a single year

First fossils of Alien multicellular animals (five times faster than Earth's)

3.45 Ga

April
March
May
February
June
January
2.5 Ga
4.6 Ga
July
December
0 Ga
August
November
September
October

First fossils of Earth's multicellular animals

1.25 Ga

The chart depicts the earth's biological developmental speed vs a hypothetical alien world's biological developmental speed, each vs. the planet's geological age, where the alien biological development is five times faster (their multi-cellular life occurring 5 times sooner (in their planet's 'February' as opposed to

Earth's 'November') - which would be derived from fossil evidence).

One could make the logical leap that the relative rates in biological development would also apply to the relative rates of intelligence development, where, given the same age in planets, the alien's intelligence would develop 5 times sooner, and thus be concurrently five times faster than that on earth.

This gives us several scenarios upon an encounter:

Encounter Scenarios
1. We could, at our present intelligence speed (cognitive speed) encounter aliens whose home planet is the SAME AGE as earth's, and therefore their cognitive speed would be faster (five times in this case) than ours, and, if they were aggressive, meaning still in early Brain-Age military conquest/enslavement mode, we would not stand a chance against them.

2. Worse, as well as the biology and intelligence developing five times faster, their alien home planet could be ten times OLDER, giving a net cognitive speed fifty times faster than ours. We would seem like mental sloths to them. Let's hope they help us across the road.

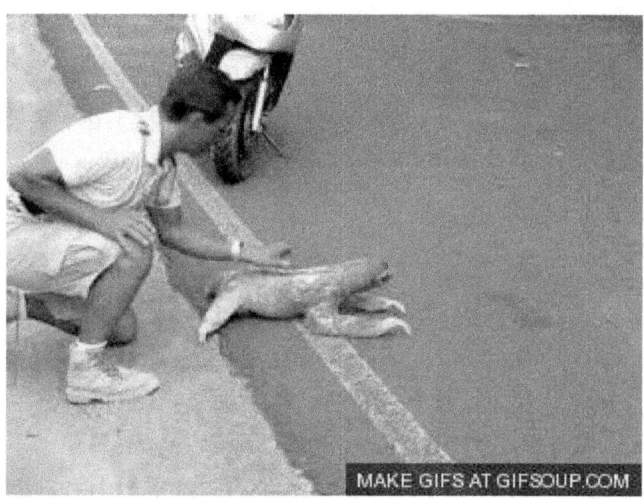
man helping a sloth cross a street safely

3. Better for us would be meeting them at 'cognitive par' - where their planet is only one fifth as old as earth, and thus their biological development only one fifth of ours, making their cognitive speed equal to ours. As you can imagine, the odds are slim of cognitive parity happening, given the endless number of other possibilities.

Who Initiated the Meeting?
As for relative cognitive speeds, whose is faster may be indicated by who initiating the meeting - where (you would think) the initiator would have the upper hand, being the initiator - whom you would assume would have the higher cognitive speed...

but... (there is often a 'but')...

On Curious Insects and Obnoxious People

Consider our earthly insects compared to us, where, some insects tanking an affinity toward humans, 'initiates' a 'meeting' with us. This does not make them cognitively superior, just curious (or hungry); and what is our usual response? Do we appreciate this affinity? Do we wonder if this particular insect possesses unusual curious intelligence over the rest of its species? No! We try to squish it! Or at least we try to shoo it away, or create more distance between us. Rarely do we want its company. Worse, toward such 'bothersome' insects, we often wish we could eradicate the entire species on the spot.

This bodes ill for the human race, with all the obnoxious people who still exist among us. As with the sloth, our only saving grace would be to be at least a little 'likable' - maybe even adorable (or at least creating an undesirable mess if we are squished), but with obnoxious people among us, we would be bothersome, and either shooed away or squished, or our entire species eradicated as a result.

The question here is, shall we eradicate obnoxious people ourselves as a precautionary measure, or just their obnoxiousness? What we are able to do depends on our current skill level (which, quite frankly, sucks, and which bodes ill for the obnoxious person's general continued existence - for the stronger the social skills one has, the less extreme one has to react).

Lucky for obnoxious people, something speaks well for our presently living in the dark - where we would never consider this, or thus eradicating them, thus sparing us from yet another episode of mass slaughter in history (which illustrates our need to develop our social skills (progress further into our Brain Age), where we then can then proactively recognize, and then deal with (in a less extreme manner) in advance such ill aspects of our nature before they actually cause us harm.

This brings us back to the Brain Age, and specifically where we stand in it (and I would say we are still in our infancy). We also need a New World Philosophy to guide us further into our specie's Brain Age, rather than relying on make-believe (as we still do) or views based on ignorance, which are inadequate, since they are not based on reality. Evil will not either, since it has the wrong values (destructive to life).

Journal 127: Survival vs. Proactivity

Consider this Chart:

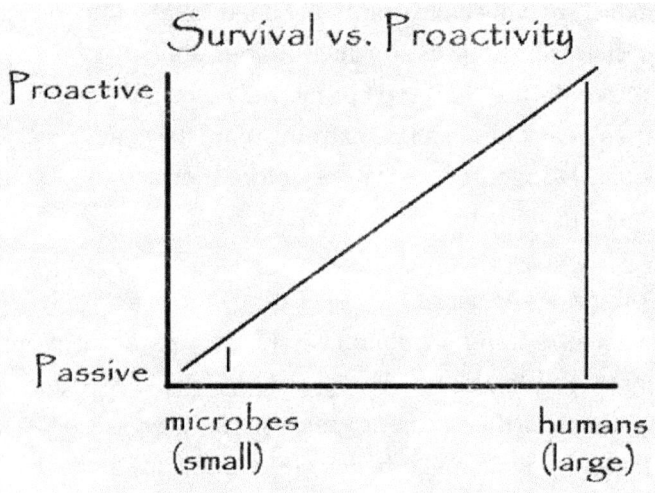

What this chart is meant to show is the 'class of actions' that are required to survive for microbes (small living entities) and humans (huge living entities - if taken as a single unit, rather than as a colony of microbes and cells - and we must).

Microbes
Microbes, having no higher cognition, as a result are defined as 'passive beings' - meaning they cannot

discover and take proactive steps against external cosmic threats to life, such as killer asteroids, while humans can. The three main strategies of microbes are all 'defensive' - procreation (numbers), dispersal, and variety.

What these three strategies mean is that microbes are more likely to survive a killer asteroid, having dispersed themselves over a wider range than humans and in many more varieties, and in huge numbers. An example would be microbial extremophiles that live in what would be fatal environments to humans, meaning they are more likely to survive a cosmic calamity such as a killer asteroid or a planetary calamity such as a flood basalt episode, while humans still exist precariously on the edge - living in an extremely limited range of fragile environments by comparison - microbes being able to live deep beneath the earth and deep within the oceans - both providing more insulation from such calamities.

Humans
In order for humans (large entities) to survive, they cannot live passively (for example, by not using their brains, or placing faith in make-believe (which includes all religions)) - they need to face reality, which entails discovering and take proactive measures against the threats to life that the harsh universe contains - not only for their own species, and not only to protect the higher consciousness that they represent (and which

took billions of years to produce), but for all of life - as if in a guardian role, which we are just awakening to, and which we will more completely embrace (and gain skills in) as we progress into our specie's Brain Age.

One 'defensive strategy' would be to reduce our size, thereby being more able to survive using microbial strategies - that of sheer procreation, adaptive variety ('diversity' to us), and dispersion. The potentially fatal flaw in the microbe's strategy is that their dispersal is limited to this planet, although they do have the ability to hitch rides on space debris, but in very limited, simplified forms; another problem with this is time - lower lifeforms may not have eternity - considering that this 'Age of Stars' will not last forever.

A problem with size reduction is competitiveness with aggressive cosmic beings who are just entering their Brain Age (meaning they do not yet see (or have the strength to allow) the value of life other than their own) - though with smaller size comes larger numbers, where we could theoretically employ virus tactics against larger, malevolent beings.

Another strategy would be to seek more insular places to exist, but this would detract from our ability to discover and take proactive measures against cosmic calamities - in effect it would be a 'defensive measure' (passive) rather than an 'offensive measure' (proactive).

Conclusion

If we wish to remain ultra-large beings, or if we wish to survive until more practical smaller physical forms are devised, we need to be proactive against all the threats that the universe can throw at us - we cannot live passively like lower lifeforms - we do not have that luxury, life does not have that luxury, which includes the widely-dispersed, many-varied, highly-numbered microbes.

Journal 126: A New Hypothesis on Biological Death

So I've provided answers to all the major philosophical questions already, such as "Where did we come from?" "Why are we here?" "Where we are going?" "Is there a God?" not to mention just plain 'Why?' and even worse (or at the pinnacle of the Philosophical Question Pyramid) 'Why Bother?" (which are all summed-up in this cute little piece on future Artificial Intelligence: http://allpoetry.com/journal/11883666-120-Artificial-Intelligence---Future-Scenarios-by-Numi-Who)

Now it is time to turn my attention to Biological Death (which I shall assume, based on observation, takes our consciousness down with it) (disassembling it and dispersing the atoms, which then say, "Well, it was fun being him; now what?" and off they go, looking for another assemblage to be a part of).

Thinking About Death in the Conventional Sense
Thinking about death in a conventional sense, we would assume that our decay occurs from within - that there is something inherently wrong with the way we are put together that prevents us from continuing on indefinitely. Scientifically speaking, one would say that the cells which do not regenerate - particularly the

78

nervous system, gradually 'die out' and cease to function - which includes keeping the heart beating and the organs and immune system functioning.

So, conventionally thinking, it is our non-regenerative cells that die, eventually snuffing-out our consciousness, which is stored in, and which emanates from, our brain cells, which quickly die after the heart stops beating, taking all your internal information with them.

Another theory takes the defective genetic reproduction route - where the 'making copies of copies' gradually degrades the genes, until the genetic code becomes so dysfunctional, we die. This seems strange, since such degradation would be random, and not the same, as it is, for all of life (for example, apples and humans wrinkling with old age - you would expect varying symptoms even within one species, given the random nature of gene degradation due to faulty copying).

Note that this does not take into account internal attacks - such as by competitors and parasites on microbial scales, such as from rogue bacteria and viruses - and I am not including 'sickness' - which can be (and has been) battled, but the phenomenon of 'growing old and dying', which we still have no clue as to the reason or cause.

Yet another non-radiation possibility is pre-programming - we are pre-programmed to gradually weaken, decay, wrinkle, and die, the solution here being reprogramming, most likely on the cellular level.

My innovative, analytical mind did not fail me, and I have identified an alternate possibility for the cause of death in general - one that may point to a new path of scientific inquiry and testing.

But first, picture this scenario in your mind: All these life forms sprouting up from the earth, only to be beaten back down again. The forms that survive are those that reproduce, which gives new generations a chance at immortality; and so far, all life has failed at it.

Why?

I've already given you the conventional wisdom (I am not going get into make-believe - pretending make-believe has, or make-believers have, anything of value to offer - and 'make-believe' includes all religions).

Now I am going to give a new (and plausible) hypothesis as to the cause of the death of all life as we know it.

Ready? I am going to extend, with inductive reasoning, what we currently know in science, and project it into the unknown, which is far different from simply fabricating fantastical claims (i.e. 'making things up') that are not projections of anything real.

An Alternate Hypothesis for Death and Decay

Rather than looking for 'internal' causes (though there are many in the 'sickness' department), and spending all of our efforts on remedying those, perhaps death and decay are caused externally. What I am primarily thinking of are the various forms of radiation that the universe bombards us with. We know that too much radiation kills cells; well, extending this fact, it is not hard to imagine the low but steady rain of radiation hitting us from space (sources which include our sun), gradually wearing our non-regenerative cells down, until the day when there are no longer enough to offer us life support.

This is experimentally testable - the challenge is finding a radiation-free location for the test group (the control group would live normal lives on the surface of the earth).

I considered deep within the earth, which is free from most forms of cosmic radiation, barring neutrinos, but there are a lot of radioactive metals within the earth which would contaminate the experiment.

At any rate, it is an avenue of inquiry that has not been done yet (primarily because an innovative analytical thinker has not come along and postulated it yet).

Is This Really an Innovative Hypothesis?

If it has been postulated and tested already, then that makes me one of those country bumpkin too distant from the centers of activity to realize his theory has already been proposed and explored via coordinated thought and research; but at least I am an original thinking out-of-touch country bumpkin - if only in parallel with those who have had such notions before.

It was difficult in the pre-Internet days to research current work on a topic, but now the Internet is maturing and is somewhat developed, and I can at least scan the headlines for work in the area... so let me scan... radiation is held not to be a major factor in our health. It is admitted that such cosmic radiation (from the sun in this case) is the primary hurdle in interplanetary travel (which I would assume includes interstellar travel, since it would take time to distance oneself from the deadly radiation of the sun; but there is no mention of such constant (if low) radiation is the primary cause of death on our planet.

Conclusion

So it could be that we already have the 'stuff of immortality' - it is just that we are being beat down by radiation (or some external cause that we haven't considered yet). I did not count sickness, since healthy people also die of old age. So the plan of action would be to consider the possibility in light of what we know, and in light of what we may not yet know; and then, if deemed worthwhile, begin new avenues of scientific investigation into the matter - meaning devising tests to verify or disprove it.

As for simply 'extending' our lives in the meantime, I have had an original thought on that: _http://allpoetry.com/journal/11693066-87-Doubling-Our-Conscious-Lifespan-by-Numi-Who_

Journal 124: Origin of Life - Insight

I've described the chaos nature of the universe as a 'chaos system' and how life operates in it (see "Chaos Theory for Microbiology"), but I did not probe the 'origin' of 'life'. Well, I had an insight today, and it all revolves around left-right symmetry...

Consider the universe. First we had a chaotic universe, then matter, then random attraction between bits of matter.

It has been noted that matter in the universe has a tendency to 'self-assemble' into structural forms, a snowflake being used as an example.

Now consider proteins - extremely complex strings of amino acids, and consider RNA, the duplicators, and you can see that 'protein duplication' is THE critical factor for 'life'.

Now apply this self-assembly tendency of matter to the 'origin of life'. First imagine molecules randomly self-assembling into amino acids, and then amino acids randomly self-assembling into proteins - and I say randomly, because the universe is a chaos system - and I don't care what the odds against such assembling is (which is one argument against 'randomness') - it happened, thus it has beat the odds. So if our universe tends toward a certain 'self-assembly order' that increased the odds, great.

Now most proteins (and we are talking about trillions upon trillions over billions of years) assembled, sat there for a while, and then disassembled (why - that is another issue, and I did cover it in the Microbiology journal mentioned above).

Then one day, one came along that 'sparked' life... How? That is the insight. Let's say we have all these amino acid strings self-assembling into random proteins, and then one day (let's say around 4 p.m.) a string assembled that was symmetrical in form - meaning it had identical left and right halves. So it sat around a while (maybe a minute), not doing anything in particular, and when it came time to 'disassemble', it split down the middle into two identical halves. Now 'life' is half-way there.

Now imagine each half, just by the way it is chance structured, 'attracting' new material in a way that it becomes a 'whole' again. Remember, trillions upon trillions of others tried it and failed - they did not enter into a cyclical process. So now we have a symmetrical whole again, and we can imagine the process repeating - separating, attracting new material, and becoming two halves composing a whole again; so, willi-nilli, you have 'multiplying' entities everywhere - which will continue multiplying as long as there is enough new matter around to attract. We define such multiplying entities as 'successful', and the surrounding available matter as 'the environment', and the method of attracting new material as 'eating', or, in a more active sense, 'hunting' and 'fishing' and 'farming' and 'gathering'.

Note that the ability to multiply alone does not guarantee life - there are plenty of examples of 'multiplication' going on in inanimate matter - the above snowflakes are one; crystal growth is another.

Now imagine, over eons of time, where a variety of successful little entities arise and 'evolve' (by pure chance). Also, in this scenario - imagine that the more successful entities happened upon better ways to 'attract new material' - meaning they became more active in the endeavor - some developing (by pure chance) limbs for mobility, and mouths and teeth that ensured a competitive edge - becoming more adapted to and successful at the task of gaining new material, and they became more complex, developing new senses and abilities, and some became much larger in size.

Such growth in size and complexity would include new combined coordination among previously independent and separate entities, giving rise to things like our different organs, which all have different functions, but which all contribute to our 'whole', in our case, our existing as a mega-entities. So you can now imagine 'us' as merely one of the mega-entity results (and our "meganess" may be a handicap - for it is a lot to maintain, just to note that).

So here we are, lumbering around as super-mega-entities, and what evidence do we have of this 'left-right' symmetry. Well! Just look in the mirror!

What this means is that, in a simplified sense, if we we could asexually divide like a cell, we would divide right down the middle - our left half and our identical right halves going their separate ways, that then gradually attracting new matter (assuming that the half-mouths could still 'attract' (or 'acquire' in our case)) new matter, which, just by the chance way our halves attract new material (based on molecular structures), results in two new 'wholes' with two identical halves each - the original and the new one.

Unfortunately for us, many of our organs did not cooperate in this left-right symmetry, so we cannot self-divide down the middle like a strand of DNA, which is why we must do our dividing at the molecular level.

So there you have it - molecules (proteins) that happened to have form in left-right symmetry, and that happened to have disassembled down the middle, each half of which happened to attract new material in a way that two new left-right symmetry's formed, and the whole process repeating, and through sheer randomness and time, eventually spawned us, and, by miraculous chance, higher consciousness (which most of us completely waste - having no guiding philosophy to grab on to (other than my New World Philosophy which is untested yet.

So happy dividing - though, funny for us - we chose the easy route - rather than relying on the time-consuming process of finding new matter randomly and rebuilding our new half from scratch, we go out and find completely assembled 'new halves' (contained in a person of the opposite sex) - which makes life interesting, and easier, and more practical - we don't have to eat one another - so it is a choice not just of convenience, but it increases the odds of the perpetuation, and thus survival, of highly-complex, largely left-right symmetrical, entities (who largely waste their conscious time)...

and happy entitying - in your chaotic (and time-wasting) way.

Journal 123: A New Dimension On, and a New Condition on, Relative Time

Einstein's Relativity

Einstein discovered that gravity and speed will affect the perception of the change in matter and energy between two observers in space, rendering time relative, if the timekeeping system spans differing values of gravity and speed. Here I will explore the effect of size.

When man first began to wonder about time, it appeared to be a philosophical issue. Before science, it was - for reasoning and logic alone were used to describe the mechanics of the physical world; unfortunately, before the philosophy that held that the scientific method was the only way to really discover reality, the philosophical notions were nebulous (metaphysical), and erroneous (metaphysical), though entertaining and colorful (metaphysical), which has led many people to nebulous, vague concepts of time, even to the point of attributing a physical property to time (and, in extreme cases, a persona), rather than correctly viewing it as being an agreed upon unit of chronological measurement (via chronometers - or 'timepieces') -

and I say 'relative' in this sense as being the change of matter and energy in relation to the change of the matter and energy of the regularly oscillating mechanism defining the timepiece (the more 'regular' the oscillation, the more useful the timepiece in coordinating activities).

We give time 'units' via a regularly oscillating mechanism (found within our timepiece (our 'chronometer'). The more regular the oscillations of the chosen oscillating mechanism and the more easily mass-produced the oscillating mechanism is, the more useful the mechanism will be to an entire population recording and coordinating their activities around time.

A coordination problem arises when this population (hence their collective timepieces) span varying gravitational fields and relative speeds, where timepieces then run at different rates, desynchronizing the timepieces, where corrections are then needed to keep them synchronized (which can be done in advance - if all the instantaneous intensities of the varying surrounding gravitational fields and speeds of the coordinating people can be predicted (over the selected agreed upon time period)).

So science has taken over the role of defining time, and, since Einstein, the 'relative' nature of time has been expanded beyond the mere rate of change of external matter and energy in 'relation' to that of your local timepiece, to the relative rates of change between two timepieces spanning different gravities and speeds, and, introduced here, sizes.

The Definition of Time - Time and Change

As long as there is 'change' in matter or energy, meaning just one unit of matter or energy in the universe, then there is something for us to apply 'time' to (and better if its units are systematized and agreed upon).

It is that simple. If a unit of matter or energy is changing, then we have something to apply time to. When matter and energy stop - say if it all falls into a singularity and is frozen there, then there will be no need for time - ever again. Lucky for us this has not happened yet (and we are the reasonable proof).

So if there is no change, there is no time. Note that time merely takes a break until matter or energy changes again; but if everything in the universe becomes frozen - meaning no more change in matter or energy, then time will take a permanent vacation.

As long as there are two singularities, however, then time exists - for they are 'changing' 'relative' to one another - in position. When they collide (which is a certainty), then all hell will breaks loose again, and there will be a lot of changing matter and energy to apply time to.

Considering the above, we can place a measure on how 'quiet' a particular area of the infinite universe is by what proportion of matter and energy is bound-up in singularities.

Let's test this 'time/change' idea against gravity and speed within a Big Bang's singularity. Einstein's equations show that the relative change in matter and energy between two objects varies relative to the gravitational fields they are in and to their relative speeds.

On Relative Time and Travel at Cosmic Distances

Curious then how, at the speed of light, our mass would be infinite, yet the mass of photons (units of energy), which travel at the speed of light, are zero, meaning 'energy' may be the way to go when traveling great distances - if one wants to avoid the infinite mass barrier, and the passage of time itself - for time would stop for you at the speed of light - which brings up another conundrum - if photons move at the speed of light, their movement through space should be instantaneous, relative to us, since our perceived time of their change would have slowed to a standstill). I'm not sure if I have it worded right, but the result would be the ability to bounce around the universe at the speed of light and lose no time.

Back to the Big Bang's Singularity

Let's consider time at the center of the Big Bang's singularity. Time speeds up due to a lack of speed (which is taking the opposite view of the equation - where traditionally we say time slows down as speed increases). in fact, everything is so mashed together, matter and energy can hardly move, and time (relative to this matter and energy) speeds up phenomenally - trillions of years can go by 'in an instant' for the matter and energy - hence 'time speeds up' for that matter and energy. Again, if all the matter and energy in the universe is in that singularity, then this time differential applies to 'everything'. Time cannot 'stop' altogether given infinity and eternity (meaning the singularity cannot 'capture everything'), which is lucky for us, otherwise 'everything' would eventually coalesce and freeze, and, there no longer being any change in matter or energy, time would stop - permanently. You might consider the positional drift of the singularity as 'change', but no - positional change is relative to an external point - which, lacking matter and energy to form it, there will not be. This is why cosmologists say there was no space outside of the Big

Bang's singularity - there was no matter or energy out there to form it; yet, giving infinity (which gives rise to an endless amount of mass and energy), this presumption would be wrong - unless it would be inevitable that all matter and energy coalesced into one singularity - then our only saving grace was that the singularity remained unstable (something inside remained in motion or in a changing state), and it was not completely frozen in place and state, and then only if that instability eventually resulted in enough perturbation to cause an expansion (an ejection of energy - at least by a percentage of the total energy contained in the singularity).

So let's assume hypothetically (ignoring infinity and an endless amount of matter and energy) that the singularity eventually attracted ALL matter in the universe (which would be a certainty given a lack of infinity and endless mass and energy) - where all matter and energy are now so packed under gravity that no matter can move (change position) (if any still existed) and no energy could change state - then time stands still - forever, until something external exists that can 'change' it again (perturb it) - which, in this case, there will never be.

So no 'change', no 'time'. Talking about time as an independent entity (as the Buddhists do) is silly; and saying time is an illusion, or has multiple (or mystical) properties, is mental foolishness. Time is a measuring system we apply to 'change'. If you want to apply mysticism to anything, do it to 'change', not the system we measure, record, and keep track of it with.

Needless to say, relational phrases like 'time is money' should not be taken literally - they depict relationships between change and outcomes.

The Flow of Time
As for time moving backwards - since time is merely an observation of change, even if the 'change' were to reverse, time would still be going forward.

Summary on Time, Change, and Einstein's Relativity

Time and Change
As long as there is 'change' in matter or energy, there can be 'time'. If all matter and energy cease to change, time will end.

Relativity
If a time system spans different gravitation fields and different speeds, then the rates of change between timepieces will vary relative to one another, which must be corrected to keep the timepieces synchronized.

Introducing Relative Time's 'New Dimension' - Size

In my exploration of micro-galaxies, I extended the fact that small systems operate faster than larger systems - a shorter pendulum arm will oscillate faster than a longer pendulum arm; the heartbeats of smaller mammals beat faster than the larger; the legs of smaller mammals 'oscillate' faster than the legs of larger mammals - you can see that this refers to relationships between 'like' systems - in these examples pendulums and mammals. I extended that to the movement and rotations of galaxies - smaller galaxies will move and rotate faster, and taking size to the extreme, meaning micro-galaxies, their speeds would be phenomenal in relation to ours. Extending this to similar lifeforms - say mammals, the mammals in those micro-galaxies would have lifespans in our sub-nanoseconds, making communication with them a challenge, and beyond us at present.

Considering this, it can be seen that the timepieces involved (ours and those in a micro-galaxy) would run at different speeds - for example, say both of our timepieces used an oscillating mechanism that was the length of our bodies. The mechanism in the micro-galaxy, being that much 'shorter' than ours, would oscillate that much faster, and hence their 'time' would be faster - they would chronicle it faster - several billion or trillion of their oscillations would span one of ours.

So, just as we have time zones to allow the sunrise for everyone on earth to occur at a local 6 am (roughly) - meaning we are changing local times to keep an external event occurring at same time of day for all the timekeepers around the planet, so too would we change local times to keep events relatively the same time span, but instead of varying locations (as with time zones) we have varying sizes (speeds). What this means is that if you were to travel from our 'size' to their 'size' (from your present size to that of the denizen of a micro-galaxy), you would adjust your timepiece accordingly (though if it changed size too, it would, in effect, adjust itself - being changed by the change is size).

The question is, is this 'change in time due to size' actual (physical) or a mere convenient time-keeping contrivance? It is a contrivance - in that it would render your timepiece impractical in the new environment; and it is physical if you brought your timepiece with you, for its oscillating mechanism would have changed length in relation to the timepieces you left behind.

Introducing Relative Time's New Condition - Similar Objects

So gravity, speed, and size affect relative time if the time system spans varying gravitational fields, speeds, and sizes (and here I introduce the newly discovered condition) - for like objects - and I think 'like objects' is a critical factor here - an additional condition - because between differing objects, there would be no perceived differences in time - whatever occurred would be perceived as 'normal'. On the other hand, if you were perceiving an object in a different gravitational field or at a different speed, and you had a like object next to you, you COULD then perceive the difference between the two like objects.

These 'two like objects' can be the same object - where you are comparing two different (relative) 'records' of it (in this case its properties of change over time, but could include relative mass, and (newly deduced), size).

Related Thought - The Limit of Our Biological Senses

Why can't we perceive the dilation of time as we move on earth (the relative time of our change slowing as we move)? Because we did not need that perception to obtain nutrients (or to avoid being caught as such) - natural selection had no need to weed-out such abilities, or, conversely, such an ability would not have offered any advantage - and it could have been possible that some lifeform of the past DID develop such a sensory input, but, since it offered no advantage, it died out by chance.

Journal 121: Starting a Conspiracy Theory - Human Spontaneous Combustion

What's a book that doesn't introduce at least one new conspiracy theory?

Quantum Entanglement
Quantum Entanglement" gets into the 'voodoo' of quantum theory. In short, it holds that if you change the spin of an atomic particle, an entangled particle will change its spin also. The distance between particles has no bearing - they could be across the universe.

Genius
Now, they say that a sign of genius is discovering new relationships between previously unrelated phenomena (though those who say it are not geniuses - they are merely relaying what they overheard in a bar). So I, pretending I am a genius, have 'discovered' a 'peculiar relationship' hitherto unheard of (mainly because it so preposterous), and I have also extended it into a new conspiracy theory.

The Peculiar Relationship
You've heard of people spontaneously combusting,

well, they are obviously the victims of entanglement experiments being done at the other end of the universe -

The Conspiracy Theory
Or, just to start a conspiracy theory, such occurrences are the result of extraterrestrial military experiments in quantum entanglement designed to annihilate individual humans if they misbehave.

It hasn't been applied yet - so far the experiments have been random, usually hitting unlucky people as they slept in their beds. Leaked news reports have indicated that accurate targeting is only months away.

Journal 114: Fuzzy Blue Neutrons

Particle physicists have theorized that there must be a special force keeping all the mutually-repelling protons together in a nucleus, along with all the neutrons, but they haven't really determined the source of this force they've just assumed it must exist. Well, I have an atomic model that answers that question (at least a loose possibility) (and at least a colorful conjecture) - 'Fuzzy Blue Neutrons'. These fit in with my Stationary Model of the atom.

Let's begin with the neutrons. At present, the neutrons are just sitting there, not doing anything, while another question is why did they decide to congregate together, and with protons no less, if they have no attractive or repelling charges to move them around or bring them together? It make no sense - unless you consider this newest aspect of my stationary model - fuzzy blue neutrons.

Think of it! (how wonderful it would be) - The fuzzy blue neutron - one half of a Velcro-like pair, the proton as the other half. NOW the strong nuclear force has a legitimate source - a physical interconnection, and the neutrons now play a legitimate part - providing a means of sticking protons together - where their complimentary fuzziness gave us the atomic nucleus and all the varieties of elements and isotopes and wonderful forms of matter that we enjoy in our local (emanating from our own local Big Bang) universe. A proton without a fuzzy neutron stuck to it is a lonely subatomic particle indeed, meaning that lonely hydrogen isotope must be the saddest particle in the universe ('multitudes of the lonely' - to poetically reflect upon it).

Second from last point - this would explain why so much force is required to dislodge a proton or neutron from an atomic nucleus.

Last point - this would explain isotopes, and why there can be more neutrons than protons in heavy elements - as the chargeless neutrons are able to stick to different sides of protons - meaning one proton could have more than one neutron stuck to it. We could address a variety of surface areas for protons - the possibility that they could vary - for example there could be irregularities in proton shapes (as well as all atomic and subatomic particles), for example, one may be shaped like an octahedron and another like a kidney bean, offering varying degrees of available surface areas of attraction.

The Fatal Case Against the Fuzzy Blue Neutron
The fatal case against the 'fuzzy' blue neutron is that, according to current theory, if a proton loses a

neutrino, it turns into a neutron, and vice-versa. Now this does not make sense - for neutrinos are chargeless - and if a proton loses a chargeless piece of itself you would reason that the proton would keep its positive charge. So there is still a lot to understand (or misunderstand if I am missing something). Where neutrinos fit into my stationary model is yet to be determined (my mind has run out of gas at the moment).

Neutrinos shooting out of a nucleus would appear to kill my fuzzy blue neutron / proton model - though with the neutrino, it is so small in comparison to a proton and neutron that it is negligible in size, and may not be a factor against the fuzzy blue model after all - many of them may 'settle' on a nucleus as dust on a ball, and be driven off with sufficient 'breath'.

What The Fuzzy Blue Neutron Atomic Model Theory Really Illustrates
Here I've created a hypothesis that I really, really want to be true - a universe comprised of fuzzy blue neutrons (think of the wonder of it again!), and this illustrates several things: 1. That someone unaware of existing data may formulate already-discarded hypotheses; 2. That there is still not enough data available, and there is still room for creative alternate hypotheses; 3. That, even though a theory is silly, it may be so attractive that people 'will' it into

acceptance, or people wish it to be true so much that they will believe it in spite of conflicting data.

Worse case is it if the fuzzy blue neutron model is hopelessly wrong (meaning it does not answer all questions), then I need to let go of my theory - to resist clinging to it just because I am sentimentally attached to it. You see this in science all the time - pet theories that refuse to be placed in the dustbin of noble but failed theories - though in my fuzzy blue neutron's case, it could have a slightly happier ending - a somewhat brighter, and possibly profitable, consolation - that is, to be relegated to the dusty toy bin of history (plush and stuffed) and to stories of fiction and fantasy. Then again, maybe if I wish hard enough, it will actually be true...

Fuzzy Neutrons and Dark Matter

Dark matter is merely comprised of particles (such as protons an fuzzy blue neutrons) without the correct fuzziness - i.e. defective particles, that either lack fuzziness ('bare particles') or have defective fuzziness (mangy particles). This makes for a universe with a lot of misanthrope particles (just to anthropomorphize things).

Fuzziness and Radiation

The fuzzy wisps on a neutron's surface are at a state perilously close to pure energy - meaning it would not take much to completely 'vaporize' the strands, which would explain natural radiation - the constant motion of matter in the universe comprising the energy that vaporizes such fuzzy strands, and these loose strands, speeding through space, are the radiation and associated particles.

Journal 112: Tricky Travel (on a Cosmic Scale)

Let's begin with quantum physics and photon entanglement, which (it has been actually observed, or at least they think so) states that if one photon is acted upon, it affects another somewhere at a distance, as if they were 'entangled'.

This has given rise to many theories, some cosmological, such as a multitude of super-dimensional string theories and multi-universe membrane theories, all trying to accommodate and explain such entanglement.

Let's suspend our disbelief and play with these resultant theories, and the notion of quantum entanglement, as it would apply to interstellar/intergalactic/intra-mu ltiverse travel, where, rather than relying on propulsion, we tap into (if not outright control) quantum entanglements.

Envisioning this, we can imagine streaming our component atoms along a string and instantaneously appearing elsewhere; or poking through membranes to vacation in hyper-dimensions (spatial dimensions of greater than three - and here again you must suspend common sense, definitions, and disbelief).

For any non-propulsive based travel, I will (at this juncture of scientific discovery) refer to as "Tricky Travel" - sliding along entanglement strings and shooting through membranes, for example (and just to note - strings are attached to membranes -but only at one end).

Is 'tricky travel' the way of the future? Hopefully, because conventional propulsion, even with gravity assists, are woefully inadequate for interstellar travel and beyond - and it is said that it will never get us beyond the reach of our own sun (meaning beyond the Oort Cloud).

Let's do some travel distance/travel time calculations based on conventional propulsion to illustrate our need for Tricky Travel:

Pluto, for example, is roughly 39AU from Earth, which takes our current probe 11 years, while the Oort Cloud extends out to 100,000AU (roughly 2 light-years). Doing the math, it would take us, traveling at probe-speed roughly 28,000 years to leave the sun's gravitational influence, and around twice that to reach the next star, or 56,000 years (which doesn't seem like much). If we wanted to travel half-way across the galaxy, which is 50,000 light-years, it would take us, at probe-speed, 56,000 years for every 2 light-years, and we have 50,000 light-years to travel, means we have 25,000 56,000 travel spans (and hopefully a Travel Center of America at the end of each span) (who says there isn't capitalist opportunities out there?) which adds up to a 50,000 light-year trip that will take 1,400,000,000 years, or a 1.4 billion year trip to visit grandma, by which time our biological evolution will have spent itself, and we will have become extinct. I hope grandma likes fossils.

So our current propulsion technology will not serve our galactic dreams, and it doesn't make interstellar trade all that practical, either. Imagine ordering a trinket from Amazon in a galactic region 50,000 light-years away. It would take your order 50,000 years to reach them, one day to process the order and ship it, and 1.4 billion years for the UPS space truck, traveling at current probe-speed, to leave your package on your doorstep. Talk about snail-mail; and your doorstep may have been subsumed far into the earth's interior by then, or (also due to plate tectonics) will have a new GPS address on the next supercontinent, and your package will almost certainly be misdelivered to an individual from the next Brain Age species.

In conclusion, if a string or membrane theory turns out to be an actual representation of physical reality, then our first engineering interest (practical application) would be to facilitate Tricky Travel.

Journal 111: How to Go Backwards in Time

How to get time to go backwards: You just count seconds backwards.

Now getting events to go backwards is the real issue - specifically how to 'undo' events in a linear manner, such as 'undo a car accident' or 'un-die' or 'un-age', where you can then turn time around and do it all over again, only with the future known this time (assuming your memory isn't erased in the process of undoing events - more on that below).

The proper perspective is one goes back to prior state. It seems that if you want to do this, you must get the entire universe to do it, too, for consider: If you want to go back to a prior event, you need to go back to the state that the event's local environment was in, at the very least; and since the local environment is seamlessly connected to what is beyond it, you would have to have that go back, too - on and on until you reached the end of the universe.

Now we are at a key realization - given infinity, we can never reach the end of the universe (for infinity has no bounds) - so it is infinity that prevents us from rewinding an event (and ultimately, out of necessity, the universe) - meaning going back to the prior state in which the event occurred. Unfortunately for your memory, it would be erased accordingly, being a part of the same physical universe which must be precisely forced backwards along the exact path it progressed.

Now we are faced with 'force' - to effect a change upon the physical world requires force, which requires an energy source, which is a problem, for the energy would also have to go backwards. All this is impossible (except for the precise application of force aspect) - since you cannot find an energy source external to the entire universe with which you can force it backwards with; even so, the 'precision' involved would have to precisely work on all matter an energy on the smallest scale, and again, given infinity, you will never reach the 'smallest' scale - there ever being the smaller.

But wait a minute - what if a machine was devised that only reversed what was inside it - say a human body? This is still a complex task far beyond our present capabilities, but it does not require the alteration of the entire universe, and is therefore a simpler, more practical task. Perhaps only a part of one's body may be affected. There is a problem here, too, however - the action may work like a wound-up spring - the more you force matter and energy back along their paths, the more pent-up force they store - meaning as soon as you open the door to the machine, your matter/energy 'interaction' with the un-rewound universe would be violent, most likely in an acceleration aspect, and the result, if seen, would be similar to an explosion. Now remember, since I have not verified this with a mathematical model that shows the relationships between force, space, time, and acceleration, this is all mental creativity, which, if not an accurate depiction of reality, is confined to the imagination.

At any rate, in regards to traveling back in time, what we are really doing is not 'traveling back in time', but forcing physical systems backwards (to prior states) with the precise application of force, and this 'force' must include acting on energy, forcing it precisely back into any matter that had transformed into it - in other words this force must be able to make energy form back into the exact particles it came from by taking the exact path it subsequently progressed along, and in a chaos system such as our universe, that is a tall order.

So one should not say 'time' - for time is merely the counter - time is merely a chronicling system of events - giving us sequential positions on an arbitrary scale, and it gives us an arbitrary scale upon which to measure future projected sequences of events; one should rather say 'events', and include the state of the universe (or locality, if you can confine it) during the point (in a sequence of events) of the event.

Journal 110: On the Strange Uniformity of Subatomic Particles

We are told that electrons have a certain masses and charges (as do protons and neutrons), and going even smaller, sub-subatomic particles, each of which have common characteristics.

Such uniformity is not in keeping with 'nature', however (at least 'macro-nature') which is random and chaotic. Consider leaves or snowflakes or grains of sand - they are not uniform - they are all various and random.

Given infinity, both inward and outward, uniformity speaks of only three thing - that we simply cannot detect the minute differences, or such uniformity was intelligently created (let's say factory produced). (There is the current scientific explanation, which I've included in my author's notes).

In broader terms, 'factory production' means 'intelligent creation' - because subatomic particles, if they are indeed uniform, go against nature (at least 'macro' nature) - they are "products of intelligent mass production" rather than products of blind chaos (the part of nature with no consciousness).

Scenarios - Factory-Produced

If our atoms were mass-produced by intelligent life out there, then how came the particles to be here? One conjecture is that we are a garden - sprinkled here and allowed to grow, perhaps even encouraged, like watering flowers, which could also be a form of procreation - sprinkling all those atoms in one place (our universe), and letting them grow, until a new permanent consciousness arrives.

Again, as I have stated elsewhere, the question is not whether a possible scenario is possible (for infinity and eternity render any probability, however improbable, a certainty), but whether or not this possibility applies to us (meaning which possibility we actually exist in - for there can only be one for 'us').

Also, as I have state elsewhere, such a 'gardener' would not use the term 'god' as a self-description, and we can take ourselves as prime examples, we who are able to create. Also this 'gardener' scenario depicts an intelligent creator larger in size than us, while my nested universe hypothesis depicts an aggregate of sub-atomic creators who have, as a whole, achieved the required level of enlightenment.

Delving further into the 'factory' notion, one would ask

if we are the product of union labor, or child labor, or forced labor, or complete automation.

Then we would ask what happened between the factory floor and us? Were we boxed or bottled? Did UPS or Fed-Ex deliver us? Were we sitting on the shelf in the garden shop, waiting to be purchased? If so, were we purchased or on sale, or were we merely a free giveaway in an unrelated promotion? Were we then planted like flowers in a garden? Are we someone's garden, and if so, just what is it that is being 'grown'? Is it our ultimate end, after all this evolution and death, to finally achieve a permanent conscious state - to finally take our place among the garden shop shoppers?

I do have an answer - consciousness, and ultimately, permanent consciousness (we are not there yet) IS nature's ultimate goal, though nature itself is consciousness's worst enemy, considering all the ways it conspires to snuff-out life (and perhaps it is an inbuilt test to weed-out sub-par consciousness's).

As it is, we are 'sputtering', given death (consciousness's do come into being, individually; but collectively, as a whole, the system still 'sputters', it is not yet permanent - it has not secured itself against its initial environment - the universe, and has not freed itself from its initial source - biology, and after that, matter and energy) (if free from them a permanent consciousness must be)..

Journal 109: Scenario Exploration - A Permanent Consciousness

So we are a species that has entered its Brain Age, and we are on the cusp of grasping cosmic concerns, such as 'who' and 'where' we are in the overall universe. We are grappling with issues such as ignorance and death on ever-broader and higher levels, to the point of leaving the biologically dependent world far behind (example - 'senses' and 'physical abilities' - our technologies have extend our biologically-based senses and abilities a million-fold).

So where do we go from here?

For me (an incessant thinker) that is an easy answer, and it is related to the issue of overcoming death - achieving a 'permanent consciousness'.

In the days of old, our primitive ancestors referred to this as 'living forever', childishly envisioning us still encased in our present human body a billion years from now. What I am saying is the focus will not be on maintaining any physical biological form (and we have already exceeded it - it no longer serves us completely, and we've come to know that it does not guarantee our survival - i.e. we've learned that we must be 'mentally proactive' in a universe largely hostile to life, especially the limited, earth-bound form we find ourselves in). Even though our intelligence IS based on genetics (meaning biology), we are at the stage (Brain Age) where our biologically-given mentality now sees the shortcomings of that biology base, and we have the mental capacity not to have to wait for biological evolution to increase our intelligence (which takes millions of years), we can do it overnight.

So it is not 'ever-lasting physical' life that is the goal. Religions have grasped this, but have only offered imaginary solutions (and yes, people still continue to kill others over which imaginary solution is preferred - though their real motive is dominance and power - the realm of the primitive mentality (still among us - we have not completely entered the Brain Age yet as a species as a whole).

The Permanent Consciousness - Explored

In "On the Uniformity of Atoms" (not yet written), I will explore a possible mechanism behind our drive for a permanent consciousness. Here, let's just content ourselves that it exists; and now let's imagine that we've just achieved it on individual levels (rather than the ability to be one combined whole - which is also a worthy goal, but which should remain at will - entering and exiting at will).

So initially, let's say we've discovered how to render our conscious permanent. In our present biological state, this would mean conquering biological decay and death - choosing a point in our biological development (such as in the physical prime or our life) to 'maintain permanently', which, by extension, will maintain our consciousness (barring sleep and being under anesthetics). Therefore, while we are still grappling with the issue of death, overcoming the need for sleep will increase our conscious time (for example, given an 8-hour sleep cycle, we will increase our conscious time from 66% to 100%, without any degradation of mental capacity (as when you stay up all night).

Let's explore this scenario. There we are, reading the news that announces "Permanent Consciousness a Reality!" What will happen next? Long, miserable lines - unless it can be purchased online.

Ah -"purchased". Will it be free to everyone? Probably not in the beginning - so only the super-rich will initially achieve it. You would think that they would come to deny it to everyone else in order to dominate them, but no - they will squander away their time in hedonistic pursuits (which is this case is a good thing for everyone else!).

After production has been in place and the production

capacity increases, the cost will lower, and the middle class will be able to achieve permanent consciousness. Think of the horrors! OK, let's not (though that is what we are here for - but I want to explore other paths other than the variety of ill-pursuits presently engaged in by the middle class (mainly vanity and envy-driven - trying to out-impress each other, which can become quite self-destructive, and destructive to the community and human spirit)).

Now let's assume the cost is so low that it is available to 'everyone' now. Everyone is happy and everything is great, right? Wrong. Now we enter the stage of 'enhancements', and the whole long, miserable line, the rich-first, repeats.

There is an ultimate limit that I see - we will find that being biologically based is a major inconvenience - our biology simply has too many demands and issues - we will see an extra-biological 'housing' for our consciousness - be it silicon or whatever. Let us assume that the best material is a new alloy. What that means is that all the matter in the universe will eventually be converted to this one alloy (in order to support all the permanent intelligences - assuming procreation is still possible, and has not ceased, once our biological state has been transcended (though this is not to say procreation will be impossible - which I also address in "On The Uniformity of Atoms").

So there is a limiting factor - the availability of 'matter' to fashion the required structure to hold our individual permanent consciousness's.

This brings up an environmental factor - that if 'element diversity'. We will lose that if all available matter is used for the special alloy. What may happen then are protected museums and parks and zoos for the remaining diverse 'original' elements. Hardest hit will be 'molecular diversity'. As it is, the 'elements' will still potentially exist in the alloy, and the alloy could be broken down to recreate the constituent elements (and I say 'recreate' rather than 'retrieve', for we will have changed the elements themselves, for example, through fusion and fission, to create such a special alloy - we would break-down an element into subatomic particles (let's say electrons, protons, and neutrons) which we will use as building blocks for the new molecule of the special alloy.

This brings up the possibility of future wars and conflicts, if the number of individual permanent consciousness's exceeds the available matter.

Would being 'energy beings' solve the issue? No - for there may be a finite amount of energy sources available (and we are right back to matter!).

Let's say that the special alloy could be made in different colors (which brings up the issue of our senses - what kind of 'senses' would we have? Would we even care about 'vision'?). What this introduces is 'fashion', and all the ills (vanity and envy) that it brings.

Back to the issue of 'senses' - that issue is a non-issue for a material housing, but what about an energy housing? Our current senses (vision, hearing, touch, smell, and taste) are all matter-based - they depend on matter (our biological bodies in this case). What of an 'energy-being'? Would it lack all senses? It could not - for that IS the definition of 'consciousness' - being able to monitor our senses (or 'inputs' in computer terms). 'Living' adds the ability to interact with the environment.

At any rate, senses or no senses, would an 'energy being' be 'alone' in the universe - unable to 'see' or contact other energy beings? It is a potential issue (to be addressed and dealt with).

What of the ennui of permanent existence? Perhaps we are the solution - being born with a blank slate and struggling to come to know our place in the universe. Perhaps it is the 'struggle' that will be ultimately valued.

Journal 93: Theory On Brain Evolution

As with anything evolutionarily biological, the brain began as a successful solution to nutrient-finding: in tandem with the development of the senses, which themselves developed by chance, and continued on due to the higher success rate of finding (and catching - given new chance motor movements) nutrients.

The senses may have initially been wired directly to responses, which, although speedier, is a less flexible system than having an intermediary decision-maker like a brain.

So the sensory inputs were eventually routed through a central 'program' - contained in the brain, which made decisions based on stored memory; the output side of the brain now controls responses and motor functions, while introducing higher intelligence, which gives the organism a competitive edge in finding and recognizing nutrients.

The more flexible the brain is, the more successful it will be competitively in securing nutrients (over creatures with non-brain controlled responses that cannot make fine differentiations).

Additional functions like greater memory, logic, and sense of self added to the success of the 'brain system'.

So, in summary, the brain is a flexible intermediary between inputs and output, giving us deliberative response.

We see this in technology today - in Programmable Logic Controllers in industry - where sensors were once hard-wired to outputs, then, in a technological upgrade, routed instead through a computer, which analyzed, compared, and combined inputs in order to better control outputs.

Here is a depiction of the PLC system - before and after:

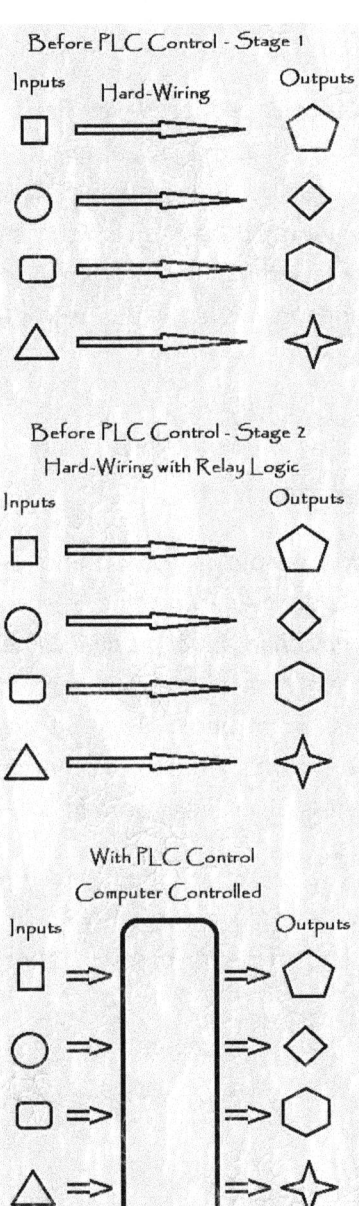

What computer control offered was 'flexibility' (after implementation) - to alter input-output relationships one merely has to alter programming lines of code (analogous to our mind-frames), which is much easier and less time-consuming than performing actual hard rewiring (analogous to altering our nerves and brain cells).

Chaos to Non-Chaos

In the beginning, encountering nutrients was a passive endeavor dependent solely on chance encounters driven directly by chaos itself. As chance developments occurred, encountering nutrients became more forceful and less a matter of chance. Senses and intelligence both contributed to this more proactive, and so far successful, willed system of nutrient encounter.

On 'willed' this pertains to our free-will - initially developed to find nutrients (based on our sensory inputs).

Journal 91: Intelligent Evolution vs Natural Evolution

Artificial Life - Living Without Natural Evolution

I was dealing with the ideal physical 'space form' for 'higher intelligence', and I said 'higher intelligence' and not 'life', because the critical factor in space habitation is 'higher intelligence', and not biology.

Backing up from there, I define 'natural evolution' as being chaos-based - there is no guiding hand making biological improvements to living entities, only chance cause and effect. I define 'intelligent evolution' as those improvements made, biological or mechanical, by existing intelligent life ('existing' as opposed to that which we merely speculate about, such as gods).

Examining natural biological evolution (below that of higher intelligence), it is microorganisms, and not larger lifeforms, that have a better chance of surviving life-opposing calamities for several reasons, all due to their small size: The relative ease of forming out of non-living matter, the requirement for less nutrients, their faster evolutionary diversity due to faster reproduction rates, which also gives them greater numbers, all these are factors which give microorganisms an advantage over larger lifeforms.

In evolutionary terms, intelligence evolution leaves natural evolution in the dust - it is far, far faster. Consider how fast our technology-based 'senses' and physical abilities have been enhanced (progressed) ('evolved') during the industrial age alone (and we can just use the last one-hundred years as a time frame) as compared to the evolution of our biological senses in that same time one-hundred year period - our biological senses have not changed much, if at all, certainly in no detectable way, in the last one-hundred years; in fact, they may have actually regressed, on the average, as our average dependence on technology grows.

What this means is that we can leave natural evolution to the eons of earth (or planetary eons, if the planet were not earth), since the natural evolutionary path to higher intelligence takes so long (on the order of billions of years if beginning with microbes). So we will not miss biological evolution, unless...

Unless our brains are too limited to perceive, or conceive of, important elements of the universe - things we need to know for higher-intelligence to survive in the long-term. Let's face it, a brain that evolves beyond what we have now would be superior

in its cognitive abilities -
but now we are back to 'evolution', and we have just seen that intelligent evolution far outpaces biological evolution, and it could be that 'technology' may give us those unforeseen senses and cognitive abilities that we would need to perceive, or conceive of, threats (and benefits - I've been overlooking those) that are beyond us right now.
Perceive definitely - that has already been happening ever since the invention of our climbing abilities (using a tree to see farther); and if you include physical issues, you would include sticks, clubs, clothes, and fire.

Cognition vs. Free-Cognition in a Machine

Cognition is trickier - for even cognitive machines are programmed by human brains, which may limit their cognitive abilities, unless we can program-in 'free' cognitive abilities - free of the limits of their initial programming.

Journal 89: An Original Clarifying Clarification Which Clarifies the Unclarified Nature of Natural Selection

Let's assume that we evolved from microbes. We see, through microscopes, how microbes attained, and use, mobility in order to seek-out nutrients (which includes finding static forms of nutrients, such as minerals, and chasing-down and eating other mobile microbes – what a great all-in-one source of nutrients – that of another living entity). We know, by inference, that they must be using 'sense' to detect such nutrients. We can conclude that our senses developed for the sole purpose of finding nutrients in an active, rather than passive, manner.

Enter sex. This complicated things – for now our senses are distracted from obtaining nutrients to reproducing. It is a dilemma in certain birds, where females prefer a longer tail, but where a longer tail is a handicap in surviving – it is not the 'fittest' who survive in this case (birds with shorter tails who can take-off faster and maneuver better), but the prettiest who reproduce. A balance is obtained, with birds of medium-length tails. But I digress - back to our senses, or more pertinent, back to natural selection.

You can see that we have more than one mechanism at work – we have two now - the drive to obtain nutrients, and the drive for reproduction. We can add a third – social needs and pressures.

Before I present my chart, let's examine social needs and pressures. They are simple in a simple social organization, and they become complex in a complex social organization. In both, they are critical for the survival of the whole. An analogy is the human body – our cells become specialized, and yet they all work together as a whole – in circulation, respiration, digestion, mobility, senses, waste management, and hormones. The same goes for complex societies – individual people specialize, yet they all work together as a whole. What this means is that a complex social structure becomes critical in advancing through one's Brain Age, and anything that threatens the complex social structure is, obviously, a threat, and can be dealt with according, without guilt or regret.

Darwin differentiated between 'natural selection' and sexual selection, when he brought up the bird tail length dilemma, but I would classify sexual selection under natural selection, and hence the need for this chart (and the whole reason for this article) – to identify the various means of natural selection.

Now for the chart:

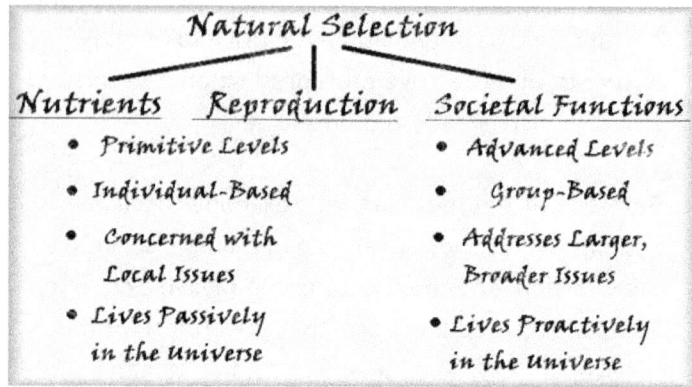

Natural Selection		
Nutrients	*Reproduction*	*Societal Functions*
• Primitive Levels		• Advanced Levels
• Individual-Based		• Group-Based
• Concerned with Local Issues		• Addresses Larger, Broader Issues
• Lives Passively in the Universe		• Lives Proactively in the Universe

You can see that, unlike Darwin, I would group 'sexual preference' under natural selection, and I would identify other components of natural selection.

On Natural Instinct

One curious observation which arose from seeing a woman familiar with large cats get out of a jeep and stand in the midst of a pride of wild cheetahs. If she stood her ground, the cheetahs gave her space. If, however, she turned and simulated flight, the cheetahs' natural programming turned-on, and they moved to take her down as prey. She would then stop and turn to stand her ground again, and they would back-off. This indicated to me that at least some of our natural instincts do not have our best interests in mind, but actually serve our predators. What this means is that we may, in part, exist, or 'survive', only as prey, to serve our predators. Lousy thought, I know...! This would be a problem - getting ourselves off the 'prey' list (hopefully without the need to eradicate our predators - for example, what a less rich world it would be without any wild cheetahs; if not rich, then diverse, which, as the new world philosophy recognizes, is still a critical component of life's perpetuation.

The Societal Factor

You can see that I deemed the 'societal functions' 'advanced', and they are. Their evolution explains our exponential progress beyond that of our simian genetically-close cousins - such progress is not biologically based, but socially-based - meaning, in our case, civilizations that allowed for individual specialization and broad cooperation and coordination between specializations, which is like hitting the evolutionary turbo-thruster switch, until we reach escape velocity from portions of natural selection, such as the need to find nutrients. See ya later, Simians.

Today, the broadest challenge is in coordinating nations. Each has a special talent, and unique abundances of resources. Our coordination can be improved; as it is, it is driven by Capitalism (centrally-controlled states have dismally failed), which has its inefficiencies and flaws, the values of which are still debatable.

So, just as biological cells specialize and group together into higher entities to address higher problems, so too with nations - we are all, deep down, the same 'cells', but we have specialized, and must coordinate to address 'higher (medical, biological, planetary, and cosmic) problems'.

A Note On the Rise and Fall of Civilizations

What has killed civilizations in the past? The cause is lack of philosophy, and an over-reliance on myth for explanations of the natural world. Are all civilizations doomed to grow and die? Perhaps they are, but not for past reasons, which we learn from, if we care to (and that care waxes and wanes).

A Note on Intelligence

You can see in the chart that I placed 'finding nutrients' and 'reproduction' in the 'less-advanced' category. This is guided by a general rule: The lower the intelligence, the more localized and immediate its concerns are.

This does not refer to potential intelligence - for two cultures can be populated by near-equally intelligent people, but in the lesser culture (the more localized and immediate in its concerns), the intelligence is 'latent' - not being used as much. It is there, it is just not being applied to broader, larger issues being discovered and faced by those applying more intelligence.

Journal 88: Ebola as Our Ally

Alien Invasion

What I will address first is the age-old situation of "common enemy"; and the first issue I am thinking of is the age-old H.G. Wells scenario of 'Alien Invasion'.

This is on the order of "War of the Worlds" by H.G. Wells, in which, as it happened, our little viral friends defeated the aliens in the end (even after the Martians walked all over Mankind proper).

So the Ebola virus would be our "little buddies" in that situation.

Evolution

Now let's consider evolutionary history - how many times do you think a new virus strain evolved and, through natural selection, weeded mankind out (along with other species) - meaning those who were not immune? Thousands? Millions?

Let's be conservative, and say mere hundreds. So that means we are walking around with hundreds of deadly viruses within us, to which we are already immune (being descended from the lucky survivors of those epidemics). Ebola would merely be the latest in that collection, and we would add just one more to that list.

Ebola will not kill everyone - there will be people with natural immunities, and their descendants will walk around, now with one hundred and one deadly viruses in them - harmless to them (meaning 'us' as a species), but deadly to any alien invader (since the conditions on earth cannot be exactly duplicated in advance by them - meaning they cannot know what they are getting themselves into in advance).

Just to note - to any alien, our planet proper will be deadly - we are fine-tuned to the current state of our planet. I say 'current' because earth's atmospheres were quite different ages past, and we, as we are now, could not have survived in them. Earth's early atmospheres had no oxygen, for example. 'Life' could exist and emerge, and it did, but it did not need oxygen (though particular varieties produced it, leading to us).

So, even on an evolutionary plane, Ebola viruses (and all subsequent 'deadly viruses') are, in effect, our 'little buddies', ultimately strengthening us against the universe.

Technology

Let's talk about 'survival' in general. Do we need to merely rely on nature's natural selection for survival? No!

We, as a species, have entered our 'Brain Age', and we can apply technology to the problem (meaning countermeasures, in whatever form they would take).

The problem here is that, living under what I'd call the "Umbrella of Technology", as compared to having natural abilities, is a weak and fleeting substitute by comparison - meaning that survival based on technology is going to be less secure than survival based on nature. For example, we cannot 'run out of' nature, but we can run out of technological supplies.

Now imagine a scenario where supplies ran out and millions of people died. We would call that a 'tragedy'; yet nature would laugh at us for being so foolish as to base our survival on technology in the first place. We would be no better than the invading aliens.

Should we then abandon technology?

No!

First, we do not have a choice - biological evolution is extremely slow and capricious. We must exist under the Umbrella of Technology and use it and take our chances - we just would not consider it such a terrible tragedy if supplies ran out, since such situations are possible.

We should at least try not to let supplies run out, however, for one basic precept of the New World Philosophy is that the Brain-Age Mind of the current reigning species is a good thing (in fact it is of paramount importance - to all of life - since it can potentially take the role of 'Guardian') - and the more minds available, the more which can apply themselves to a problem; and the more minds applied to a problem, the better the solution will be. Simple logic.

Conclusion - The Importance of A New World Philosophy

Is Ebola a problem? Yes. How many minds are applying themselves to the problem?

Not many.

Why? Because of our current feeble cultures and mind-states - our philosophies are weak - we are in a mind-state of 'leaving it to others', we are in a mind-state that does not value education, and we are in a mind-state plagued by a milieu of past prejudices and ignorance. We lack (and need) a new world philosophy.

Will Ebola kill me? Most likely - not many people will be immune to it, and I don't think I will be a lucky one. Should we let nature take its course (the preferred method - if we can afford it)?

No - only because we may not be able to afford it time-wise - meaning losing most of the human race (the current 'Brain-Age' species, though most people do not think this way yet - and hardly apply their brains to anything at all, and if they do, to trivial matters) will most likely mean a setback in technological progress, and in general (particularly scientific) knowledge, and we may not have the luxury of 'time' - to recover, or wait for the next 'Brain-Age' species. So we use the less secure method of technology to survive and do the best we can.

Epigenetics - Using Nature as Technology

They now say that what we 'do' in life affects our epigenetics, which are then inherited. So in effect (if this turns out to be true), we can manipulate what our descendants will genetically inherited, which is 'genetic engineering' (albeit through 'nature') through 'life experience' (such as stresses); and 'life experience' can be simulated (such as stresses) in order to effect the desired genetic outcome, which means we can 'speed up' our 'natural evolution', using nature itself as a tool of 'technology.

If basing life (the biological aspects, that is, which we still are not free of) on natural selection is superior than basing it on technology, then any technology that uses nature as its tool will be just as good as natural selection itself.

So there IS hope for basing survival on 'technology' - if technology uses nature as the primary effector.

Journal 87: Doubling Our Conscious Lifespan

I say 'conscious' because we spend a quarter to a third of it sleeping. If we can overcome the need for sleep, we could increase our 'conscious' state by 25-33%.

Now if we define 'being conscious' as involving activity (even if it is only 'thinking'), then we can assume that our speed and efficiency will increase, thereby increasing our actions and the results of our actions, and raising the 25-33% to at least 50%.

If we further assume that being conscious longer will increase our wisdom earlier in life, then the results of our being conscious may increase many fold.

This is all assuming that our biological lifespan remains the same. If that also increases, then the increases could be exponential.

Journal 85: Two New Charts for Life: Survivability Odds and the Need for Death

When a near-life-ending asteroid impacts the earth, all the larger forms are annihilated, and only a few microbes are left to begin the whole random process over again. One thing this tells us is that 'smaller' means a greater chance of survival against cosmic calamities. One other factor contributes to an organism's survival - intelligence. This is what a chart of this would look like:

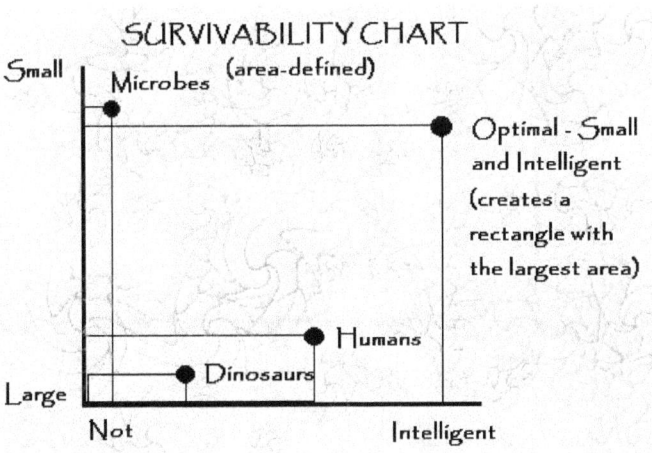

There is one other factor: longevity. Consider how 'intelligent' a dinosaur would be if it were still alive after 100 million years - would its intelligence be limited by its biology (brain size), or would it gradually learn enough about reality to develop space travel and empathy for life?

One other 'chart': It is a common opinion that Baby Boomers will linger around, growing old and burdensome, and it would be better if they just die off to make room for the next (improved, as a general rule) generation, especially if there are limited resources. While the 'burdensome' factor may be true for current Baby Boomers (for they did not grow-up in the mentally healthiest of times, culturally* speaking) it does not necessarily hold true for all generations. Once again intelligence plays a part:

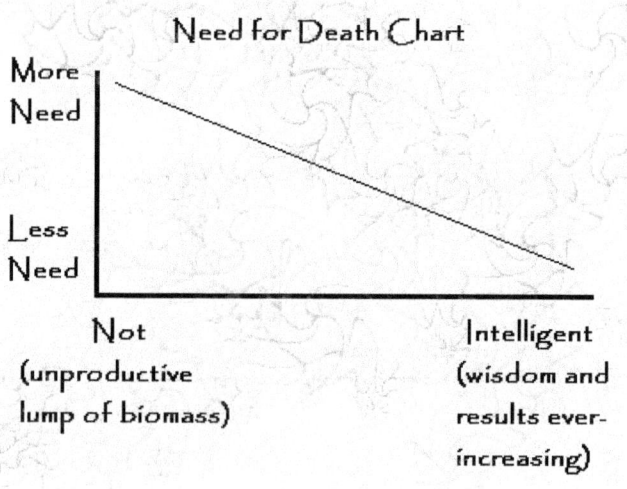

Journal 84: 'Perspective' and Cosmological Models

Let's presume that our universe will expand until it dissipates into the void, like a cloud of gas would disperse outdoors.

This means all galaxies are receding from 'us' (and let's assume 'we' will be around that long, and let's assume they've preserved earth and the sun and the Milky Way where they are today), and that the galaxies will eventually be too far away to be seen.

Now comes the 'Perspective' - let's say that, during this time, a new species has entered its Brain Age and has begun to seek serious answers (as opposed to mythical - as in 'Early Brain Age' solutions) to cosmological questions. They will have no 'receding' galaxies (no 'Red Shift') to help them formulate their place in the cosmos, for none will be visible any longer (and we are assuming due to distance, and not that they will have dissipated themselves).

So what kind of cosmological conclusions would they come up with? Most likely they will conclude that the universe contains a lot less matter than that which the original Big Bang generated, and they will not have any evidence of the Big Bang remaining.

What could change even this? If an adjacent Big-Bang-generated universe came along and crossed their path. They would discover the influx of a lot of matter (and let's say it included billions of galaxies). Based on the movement of THOSE galaxies, they may finally deduce their own origins, and may deduce that they evolved rather late in our Big Bang's history.

What can we do to help them? We can leave our knowledge behind in some time-resistant form, informing them of the Big Bang, which we deduced when most galaxies were still within a visible distance, though receding.

They could even be attaining their Brain Age status while sucked into the 'Big Crunch' of a passing universe.

Let's examine a 'Big Crunch' from an entropy perspective. Matter and energy, being entropic by nature, will tend to seek a state of rest. The problem is that matter will all tend to 'fall into' one spot, a spot that has become the 'dominant collection of matter', gravity being the attractive force, resulting in another singularity, which will, under all that stress and pressure and instability, burst apart again (go 'Bang' in a cosmic way) (or, in other words, 'rebound' outward again). I'll be in my lounge chair watching it all, though from a safe distance (and with a space heater). The species in question will first wonder where all that matter came from, and then why it is all moving toward the same point. Hopefully they will invent lounge chairs and space heaters before it is too late.

Journal 83: Traveling Between Nested Universes

This notion germinated while thinking about the relative nature of a black hole's event horizon, and how it will be further out from the singularity for a more massive secondary object due to MUTUAL gravitational attraction, and by reverse reasoning, closer-in for a smaller secondary object.

I then took the notion of the secondary object getting smaller to its extreme (why do I always do that?) and came to this realization - that, if we were to make 'ourselves' infinitely small, we could go right up to the singularity without having reached the event horizon yet (just as if, for an infinitely large object, the event horizon would be an infinite distance away).

How does traveling between nested universes fit in? Well, it would allow "us" (and I use quotes because 'we' would not travel in our usual form - we would need to 'be' very small, or we would have to stream ourselves, very-small particle by very-small particle.

At any rate, here is where the real voodoo comes in: Once we reached a certain point near the center of the singularity, it would 'propel' us the rest of the way (down in size), and if we calculated our 'shrinking speed' just right (assuming we would need to engineer the travel), we could 'emerge' into the next nested universe down.

Remember that I've postulated that nested universes come in quantum sizes - anything size between the two would have been consumed or annihilated by either the larger or the smaller universe; and remember that nested universes all share a single point.

So in effect we could only travel to the next universe down in size which shared the point at which we were reduced in size by the force of the singularity, in other words, a point also shared by the center of the singularity.

I'm not going to do the math, since I can't without more time than I have left to live - though I know it would be fun (and much more fun if it turned out to be true), and I'm not going to pretend that I would be able to understand the math even if someone did work it out. I'm also not going to claim that the conclusions would validate my hypothesis, for calculations may clearly disprove it, if something I've overlooking in fundamental physics hasn't not already. If the calculations are inconclusive, then there is still hope - based on future discoveries and technologies. At any rate, it would be advisable to determine if there are nested universes before embarking. I think the discovery (or disproving) of nested universes would come before the ability to harness the power of black holes came. The lack of nested universes, however, may not be a reason not to go through with the experiment, though where you would end up would be a mystery until you got there. Hopefully a way to get you back was also developed, or will be shortly, or at least before you cease to exist, but that is unlikely, since your time-flow will be proportionally faster - meaning if you do end-up in a micro-universe, its entire lifespan may only last a few of our microseconds; but those are mere technical details...

Feasible or not, it is clear that the force of a singularity can be used as a tool (or energy source) - if not for travel between nested universes, then for something else - and what, only future engineers may know, and hopefully they will delight in designing it - I know I would.

The Atomic-Stream Technique Conundrum

I mentioned the technique of 'streaming oneself' atom by atom through a singularity down into another quantum level universe, but that does not seem practical (even if possible) - for each of your atoms would be the size of an entire universe one quantum level down.

Nested Universes as Train Stops

Let's say one stop stops right in the middle of a singularity (a certainty if infinity exists) - this is one stop that you will then want to avoid. What this means is that 'seeing ahead' will be a critical factor; not seeing ahead will be no reason not to travel - it just increases the risks.

Journal 82: On Death, Inductive Reasoning, and Science

Consider death, or our gradual decline. The current question is, "Why?" To begin, we act on words and phrases such as "perhaps" and "maybe" and "it could be", and to solve the problem we act on phrases such as "what if..."

Inductive reasoning is used in the initial phase of "perhaps" (developing hypotheses, in scientific jargon). For example, as to death, we are still considering what investigative path would be the best to take with our limited resources (time, energy, life, and money).

Let's do some inductive reasoning on death. Perhaps it is a 'biological battery' running out, so the answer would lie in the electrical direction; or perhaps it is related to cell membranes - if they gradually change in a way that prevents their required nutrients from entering, they wither away and die. Perhaps the answer lies in groups of cells functioning together for a higher purpose, such as liver cells, and their losing that cooperative ability over time. Perhaps the answer lies on a plane well above the cellular - on a complete system level.

You can see that there are several completely different

paths to take. A robust scientific community can investigate all paths simultaneously, but how many robust scientific communities are there (not to mention coordinated)?

Inductive reasoning is never fully turned-off. For example, along an investigative path, one may inductively reason, from seeing, other possible causes, and new promising paths may arise.

Journal 80: My Thoughts on Humans and Space

We like to envision ourselves, as we are now, traveling through space, populating other worlds, discovering other life forms, but... stop the music! Right there!

Consider 'us' as we are now. Compared to amino acids, which form on, and hitch rides on, asteroids, we are HUGE. What this means is that we are not suited for space travel – we require too much energy to function, and we waste most of that energy (and we create a lot of waste – but then we would either recycle it or just eject it out into space (creating hazards for future space travelers, on the negative side, and dispersing microbes on free rides through the galaxy, on the positive).

What I am saying is that we are ill-suited (being so huge) for space travel. If intelligence on our level (and the level necessary for intentional space travel) (as opposed to passive space travel, as with our ejected microbes) can be packed into smaller biological forms, all the better – we would need less energy, and we would create less waste. We know that the 'purest' form for intelligence would be pure energy itself –

requiring no new energy and creating no waste, but ahhhhhh – there's the rub - "Requiring no new energy". Now to function, one must 'expend' energy, and it can be assumed that, even in a pure energy state, to actually 'do' anything, we would have to expend energy – so we WOULD need a constant influx of energy, albeit most likely less than our present material forms require (but who knows – it may take MORE energy to sustain a purely energy state).

So what I am really getting at is, "Quit envisioning 'us' (as we are now) as space denizens! It most likely will not happen! Or, if it does, and even our current technology can make that happen on a local scale (the moon and Mars), it will be extremely limited and localized. So if we want (or need) to travel beyond that, we should start considering 'ourselves', and what form would be best suited for 'galactic' (never mind inter-galactic!) travel. I am saying 'consider what form' since we are progressing ('technically evolving') toward being able to 'shape' ourselves into the most suitable form for a task (be it deep-sea existence in ultra-high pressures and alien environments or travel through (and maybe even existing in) the complete (and very cold, nearly absolute zero temperature) vacuum of space.

What about right now? Well, we can have (and do have) a lot of fun with the form we are in… so enjoy it! On a more serious note, our present form is the one best suited for our present environment (but note that the earth is ever-changing) – in climate and geology (which would be deadly to any visiting alien), and in biology (as in resisting all the present and past microbes and viruses, which would be deadly to any visiting alien). On earth, we are, after all, 'traveling' through space, on 'spaceship earth' as they say (and is that enough? Who knows, but if we can go beyond it, we should – it is the better bet for extended survival, though not by much, given the unknowns, and the vastness of space and time, and the perils of intelligence itself against life).

Post Note

I was listening to a sci-fi book, "2312" by Kim Stanley Robinson, and she was describing a space terrarium in a hollowed-out asteroid, around nine miles by four miles, giving it around several hundred miles of cubic space. Spinning it to create 'gravity', where the entire interior circumference could be used as 'ground', gave the interior a few hundred square miles of ground space. Now in this she had earth animals, and what occurred to me first was how inefficient and ill-suited the animals were - requiring far too many nutrients and creating far too much waste for their sustained health and well-being, not to mention the economic loss they would be if they were farm animals creating marketable goods. If the asteroid were a zoo intended as a safeguard - an extra-terrestrial preserve for animals and plants (to survive calamities on earth), then that would be different - the great expense would be justified, as it would address 'diversity' - not only in preserving old forms, but creating a whole new ecosystem where separate evolutions of life would occur as compared to that on earth.

So animals in space as an economic pursuit? No. They produce far less than they consume, and they create waste-management and a health problems, though solutions (most likely expensive) to these problems could be engineered. As decorations? They would be very expensive decorations, with very expensive solutions to the burdens they create. As pets? They would be very expensive pets, creating the same burdens requiring the same expensive solutions - animals are just too large for space, just as we are, compared to our being much smaller beings, requiring less environment, less energy, less nutrients, and creating less waste.

One issue is brain size - how much brain do we really need? Look at ants - they have motor functions and sensory inputs, and they make decisions, all on no brain proper at all - on just a ganglion of nerves. So if we only use 3% of our brains, to pick a number, then we only need a brain (hopefully with all the needed features, such as a frontal-lobe area) 3% of the size we have now, and brain size would be the deciding factor as to how small we could become.

As with everything pertaining to a Brain Age species, that species could apply technology in expanding its capabilities and reaching its goal. It could be that such a species (say us) could get by on little to no biomass if we invented suitable hardware to perform not only the our physical functions (as guided by mental functions), but our abstract thinking mental functions as well, freeing us from biology, at first in stages (we are already there - consider the car (increased mobility) the calculator (a mental process) and telecommunications (communication speed and distance). Ideally we would rely on no biomass, but this presents a problem - a lack of natural evolution.

100% Artificial Life - Living Without Natural Evolution

We are talking about the ideal 'space form' for 'higher intelligence', and I say 'higher intelligence' and not 'life', because the critical factor here is 'higher intelligence', and not biology.

As far as biology is concerned, microorganisms have a much better chance of surviving cosmic calamities for several reasons due to their small size: The relative ease of forming out of non-living matter, the requirement for less nutrients, and their faster evolutionary diversity due to faster reproduction rates, which also gives them greater numbers, are several factors that immediately come to mind.

In evolutionary terms, intelligence evolution leaves natural evolution in the dust - it is far, far faster. Consider how fast our technology-based senses have been enhanced (progressed) ('evolved') in the industrial age alone (and we can use just the last one-hundred years as a time frame) as compared to the evolution of our biological senses in that same time period - our biological senses have not changed much, if at all, certainly nothing detectable, in the last one-hundred years; in fact, they could have actually regressed on the average as our average dependence on technology grew.

What this means is that we can leave natural evolution to the eons of earth (or planetary eons, if the planet were not earth), since natural evolution to higher intelligence takes so long, on the order of billions of years if beginning with microbes. So we will not miss biological evolution, unless...

Unless our brains are too limited to perceive, or conceive of, important elements of the universe - things we need to know for higher-intelligence to survive in the long-term. Let's face it, a brain evolved beyond what we have now would be superior in its cognitive abilities - but now we are back to 'evolution', and we have just seen that intelligent evolution far outpaces biological evolution, and it could be that 'technology' may give us those unforeseen senses and cognitive abilities that we would need to perceive, or conceive of, things that are beyond us right now.

Perceive definitely - that has already been established as possible, since it has been happening ever since the invention of the wheel (and before that, if you include sticks, clubs, clothes, and fire).

Cognition vs. Free-Cognition in a Machine

Cognition is trickier - for even cognitive machines are programmed by human brains, which may limit their cognitive abilities, unless we can program-in 'free' cognitive abilities - free of the limits of their initial programming.

Journal 79: My Thoughts on Life and Entropy

The questions here are, "Is life anti-entropic?" meaning does it come into being out of nothingness; and "Or is life just riding the universe's energy wave (cause by the Big Bang) down to complete dissipation, when the universe (and life within it) will end? and, "If life IS just riding an energy wave down, can the rise of intelligence overcome its ultimate dissipation?" meaning "Can we keep creating energy (which sustains life) even though stars, galaxies, and universes (present sources of energy) 'die'?

Just a note on philosophy and science, you can see how they are driven by questions, and their main purpose is seeking answers.

So is life anti-entropic? Probably not — scientifically speaking, it does require constant energy input, and the energy must come from somewhere external to that life; and philosophically speaking, you cannot derive 'something' from nothingness (though that leaves the question of the origin of energy and matter unanswered, which is tied-up with infinity and eternity, which, as wholes, do not physically exist to us (since they are boundless) (and I say 'physically' because they do 'exist' as mental concepts, and mathematically) just as we would not exit to them (if they were 'wholes', which they can never be, and one can conclude with tongue in cheek, "Which is why we exist!").

So where was I? Oh yes, life and entropy, and energy.

Consider the Star

Let's consider energy. Cosmologically speaking, it would be nice to know the basic origin of energy. Let's consider stars. Could it be that they are the basic origin of energy, if we define them as energy sources? We tend to overlook them since they are grouped into galaxies, and galaxies into clusters – cosmology tends to project answers onto larger systems, while particle physics tends to project answers onto ever-smaller entities. So maybe the answer lies somewhere in the middle (and maybe not, but let's continue dwelling on stars).

Stars do create a 'localized' 'wave' of energy, from which life can originate (based on our own experience) and continue on, until that star dies. A side note on this – you may be thinking in a science-fiction vein, that we will grow to travel away from our star to populate the galaxy, finding ever-new stars to 'be around'. Well, one theory has it that life here, as we know it, will never even travel beyond the gravitational reach of our sun – the distances involved, and the speeds we must travel compared to what we can attain, are greater than the life of the sun, meaning we, as a species, as well as any future intelligent life on earth, will never get beyond the Oort Cloud (which theoretically extends to 50,000AU (astronomical units)(roughly the distance from the earth to the sun – 93 million miles) from the sun, which is only around 0.79 light years) (though it seems like a trivial distance when given in light years).

On another side note, one may ask, "Then how does life perpetuate itself throughout the galaxy?" (assuming it has been doing this already, and maybe jumping from galaxy to galaxy). One possibility is longevity – it simple stays where it is, in a frozen state of matter, waiting for another star, or galaxy, to come along and sweep it up.

So back to the 'star' – we can agree that it is a localized (as opposed to other stars) wave of energy (by wave I mean it has a peak, then gradually dissipates, sometimes not so gradually and rather violently, as when a wave breaks upon a rocky outcrop, or when a star goes supernova).

So we are existing on our star's 'energy wave', which will eventually dissipate (burn itself out). This means the life is probably not anti-entropic – it cannot exist without an external energy source, it cannot 'fight entropy' all by itself, though it seems to 'come into being' from inorganic molecules, meaning 'out of nothing', but really it is 'out of nothing living', meaning basic atoms, some of which, by chance, gather into amino acids, some of which, by chance, make proteins, and some of which, by chance, combine into the various components of life (and I say 'by chance since it all occurs in a chaos system, which I've postulated in another journal).

The Remaining Question

So the final question is, "Can the rise of intelligence overcome the dissipation of our localized energy wave?" Stay tuned for the answer (though it will probably not come in your lifetime, but no need to worry – there are millions of lifetimes left before the sun dissipates) (seems like a lot) (but is it enough?) (worry, worry...!).

A Note on The Universe

So the 'ultimate' source of energy for us, right now, is our universe. The question here is, "Is it inevitable that its energy content will eventually dissipate, or burn itself out?"

One reason to say 'no' is the influx of energy from other universes - adjacent to us on our own physical-size plane, and adjacent to us on the next quantum-size planes, both smaller and larger than us (I've presented my theory of 'quantum sized universes' (basically 'nested universes' sharing a single point in space) in another journal).

So yes, our universe may be burning itself out (materially speaking), or spreading out into complete dissipation, but that does not rule-out the influx of energy from other sources, which would sustain the universe (for a time proportional to the amount of influx of energy) (you can come up with the mathematical equation!).

A Note on Applying Intelligence

NOW the question is, can we just sit here passively (like animals), and HOPE that everything works out for life (though animals cannot even 'hope'), or do we need to apply our intelligence, and MAKE things work out for life...

I would suggest opting for the latter, though it isn't guaranteed that that is the path to take - considering how intelligence itself can be a danger to life (as was fully realized in the 20th Century). Perhaps being a dumb animal is the way to go - we just do not know yet; but since we do have the intelligence, and we are the only species with it, we might as well put it to use, and take that path, and see what happens (and hopefully record our results for the next intelligent species should we fail) (and we probably will - being the 'first' intelligent species around (given the lack of evidence otherwise) - for when does a first attempt at anything succeed?).

Journal 76: Probing the Nature of Reality (With A Few New Insights) via Responses to Max Tegmark

(Where I have no observation, it means I thought it was a though-provoking statement worth noting.)

New (breakthrough) insights (by me) (rather than common-sense responses or with non-original knowledge) are in italics.

Max Tegmark: A doppleganger-laden Level One multiverse in a more diverse Level Two multiverse in a quantum mechanical Level Three multiverse in a Level Four multiverse of mathematical structures.
My Observation: What is all that? That is Mr. Tegmark trying to create order and classifications of reality in order to create a mental framework from which to view the universe, and more practically, approach problems.

Max Tegmark: "We didn't even need rocket power to figure stars out, we just needed mental power."

Max Tegmark: "Prior to 1925, people did not know there were other galaxies."

My Observation: Based on a region's economic class, a lot of contemporary people in the lower economic classes (which implies less education and more grab-bag mysticism) still do not know.

Max Tegmark: "It is not enough to talk about our place in space, we also have to explore our place in time."

My Observation: *This 'exploration' is the gathering of ever-more reference points with which to 'place' ourselves against.*

Max Tegmark: "The farther out galaxies are, the younger we see them; and when we look further, we see no galaxies at all, just a plasma wall."

Max Tegmark: "Calculating speed and distance, everything was 'here' (at the same point – the Big Bang's singularity) roughly 13.7 to 13.8 billion years ago."

My Observation: That is assuming 'we' are still at the origin, at the point where the Big Bang's singularity was, and we most likely are not (probabilistically speaking, and because some of the singularity may still exist, sucking matter into it (which would annihilate us as material beings)). Science fiction notion: A cult that believes that if they are sucked into a singularity, their souls will be freed from their material bodies, and released into the universe (or another one, or the one that is Heaven).

Max Tegmark: "A super-heated gas is plasma, which is not opaque; so beyond all these galaxies is an opaque plasma screen, so it looks to us as if we were surrounded by a plasma sphere."

My Observation: This implies that 'we' are not situated near the edge of the Big Bang's expanding sphere, or the plasma screen would not look uniform in every direction – the 'surface' nearer to us would look less dense (since we are closer), and thus less bright. Consider what the universe would look like if we were at the edge of the expanding sphere - all the galaxies in one direction, and nothing in the other (unless other universes looked like pinpoints of light, like the stars we stars, or like nebulous patches, like the galaxies we see).

Max Tegmark: "We finally got some good photographs of this plasma screen in 1992, which won the Nobel Prize, and now we have even nicer pictures – which all agree - the universe when it was only 400,000 years old. So we've mapped the edge of our visible universe, is it "Game Over" for cosmology?"

My Observation: and why can we see the edge of the universe? Because it has been 'sending us light' constantly, in an unbroken stream, since it was near to us, and since it is receding away from us at less than the speed of light. This is never explained, and one ends up wondering how, so I explained it.``

Max Tegmark: "Two kinds of substances that we still do not know much about: Dark Matter (consisting of 24% of all the matter out there) and Dark Energy (69% of matter, though this is in a 'potential matter' state). The amount of dark energy will determine the fate of our universe - whether it continues expanding, or collapses in upon itself again.

Max Tegmark: "Our present 'galaxy map' is presently less than 1% of the total volume of the observable universe. Mapping the universe today is like sending out Lewis and Clark in 1804 to map out the new US continent."

My Observation: *Why was the Lewis and Clark important? Because before that, there was no knowledge of it being shared with the world – it was inhabited by primitive people who could not communicate on such a grand scale, and who in fact did not even know there was a 'grand scale', or if they had a notion that there was something beyond their bounds, it was composed of myth and religious fabrications, and was not in any way correct.*

Max Tegmark: Radio telescopes can see hydrogen waves, which are 21cm long, and longer if red-shifted.

My Observation: The visible spectrum for our eyes is 390-700nm (430-790 Terahertz). The more energy that is stored in an atom, the more the atom vibrates, resulting in a greater amount of radiation emitted. Plasma is charged ionic gas with positively-charged atoms and free electrons, which were driven off their atoms and kept off by heat energy.

Max Tegmark: "Our universe isn't just described by math, it IS math; specifically, it is a mathematical structure. Our world does not just have some mathematical properties; it fundamentally has ONLY mathematical properties."

My Observation: Math is a language, of symbols. To turn around and say the symbols ARE the reality is foolish (and such foolishness has a specific name: the reification fallacy.

Another point is that there is not a one-to-one relationship between mathematical structures and objects/events in the physical world - many different mathematical structures can apply to the same object/event on different extra-relational levels within varying degrees of context, and toward different inherent parameters of the object/event.

Yet another point is that there are many ways (languages) to 'describe' the universe, but it does not make the universe that language or description in a literal sense. For example, a table is not the 'word' table. Many forms of art can describe the universe, but it does not make the art the subject in a literal sense.

You can argue that math can predict things, but so can experience (which is why most people try to get by on experience – it is much easier than math!).

His statement that the universe has only mathematical properties is doing the word 'properties' and injustice, for it does not cover anything in the subjective world, which is a strong driving force in life. He should have said "math can describe, in only a quantitative manner, both the static and the dynamic characters of an object/event, and the inherent and relational aspects concerning the mechanisms and degrees of compositions that drive objects/events.

Another problem is defining an object/event - where do you draw the line? Meaning, where do you set your arbitrary limits? Since arbitrariness is involved, mathematical structures are inherently vague, and thus incomplete, even as descriptions, let alone as BEING the objects/events they describe.

Further, even if what he said was true, a mathematical definition is useless in a chaos system, which is exactly what life originated in, exists in, evolves in, and operates in. In a chaos system, math can only describe what is and what has been, it cannot predict the future any better than experience, for to predict a chaos system, all inputs must be known that affect the component you are studying, and you cannot know all inputs, for infinity raises its insurmountable head.

What I think Mr. Tegmark is getting at (and what is the overall value of his book) is in 'stretching our minds' - to prod us to develop new perspectives with which to approach problems. For example, If we see a spaceship as a mere collection of mathematical structures (though like I said, he is thinking too simplistically, since many structures overlap in their descriptions of a single object/event), then that may be a valuable new tool for us. As it is, the view is one of extreme reductionism, which for the most part is useless in communication things that need to be communicated, like how to drive a car - you do not begin with the quantum-mechanical state of all the car's atoms, you begin (and deal with the problem) on a much higher (composite) level.

His statement that math IS reality would be closer to the truth (or at least more easy to digest) if the universe was composed of just one elemental particle underlying all others, That would simplify things considerably, for then you could say that everything is made up of so many elements combined into whatever particular patterns and shapes (and the resultant functions) resulted; but even then the elements will not be numbers.

As for the numbers and symbols of math, a question would arise: "Which number/symbol is best suited to BE physical objects and events?" I like the number nine myself (just to illustrate the absurdity of Wegmark's claim) (sorry, "Tegmark" - meaning if he were merely trying to become famous, he failed)...

Max Tegmark: "Atomic particles do not have any physical properties at all, they are only numbers."
My Observation: That is just plain silly! (let's call it 'absurd' and add it to my "Another Generation of Absurd Birds" journal). If they have mass then they take up three-dimensional space, and therefore they will have physical properties. Without physical properties, they are pure energy, meaning they will have no mass, meaning they are energy packets, which have no physical properties. Perhaps 'physical properties' can be the defining criteria between mass and energy. Since, however, electrons are said to travel 'at the speed of light', perhaps they are massless energy packets after all, since nothing which has mass can travel 'at' the speed of light (where its mass would become infinite, and we know 'infinity' is not a 'whole' entity, and cannot be attained). This is not to say that something as small as an electron, which begins with a smaller mass and has room for a larger increase of speed, cannot travel closer to the speed of light than our human bodies can. (Note: Maybe there is something to this possibility in near-light-speed travel – addressing the problem from the perspective of individual atoms, and addressing each atom separately in the transport).

Max Tegmark: "If I'm wrong and some properties cannot be described mathematically, we'll eventually hit an insurmountable roadblock – there will be no more patterns left to discover."

My Observation: *Chaos systems are by definition patternless, and cannot be described by mathematics, at least in an exact, predictive way; math can only describe what is and what has already occurred (the history of "what is's").* As a general rule, to loosely understand and predict events in a chaos system, the more inputting "If's" we know (that have a bearing on the system under study, no matter what the level), then the more probably "then's" we can predict that much more reliably and farther out into the future, and on more levels.

Max Tegmark: "We may not be smart enough to find all the mathematical clues or understand what they mean, but they are there, waiting to be discovered, and our ability to understand the universe is only limited by our creativity and imagination."

My Observation: To 'creativity and imagination', I would add 'experience', though this is not necessary, since we can know things purely through math (though advanced in math require creativity and imagination). For example, look at the discoveries of Neptune, radio waves, and the Higgs boson - which were all mathematically 'discovered' and predicted before verified by actual 'experience').

Max Tegmark: "If consciousness can be understood as particles moving in complex patterns that process information, like inside of a computer game, then it can be described mathematically as well."

My Observation: This may very well be a good example of looking at a problem from a wrong perspective. As a physicist, he looks at things from an atomic particle perspective. In philosophy, this would be called an "ultimate reductionist" point of view. Atoms, on higher event scales, are reactionary, and provide mere structure and framework (if only for the really important entities - molecules). *For example, consider the basic mobile organism that 'eats' – it gained mobility in order to seek-out nutrients (making it more 'fit' to survive, evolutionarily speaking), and it 'kept' chance-generated senses of various types (vision, hearing, smelling, touch, taste) that enhanced the value of mobility. So you can see that the atomic particles are not playing a leading role here – there are the mere building blocks for these higher functions,* the framework; they are not even the molecules (remember, he is perceiving the phenomenon from a particle physicist's perspective - sub-atomic particles). Molecules (especially bio-polymers) play a more important role, but as collections that create higher structures (for it has been observed that when 'life molecules' group together, they begin to perform a 'higher function' that individually they could not have, such as a liver as a unit performing new 'liver functions', as compared to a collection of independent, individual cells still behaving independently). In effect, a particle physicist's point of view will miss the larger picture – the larger cause-effect relationships that can only exist on planes and complexities (and patterns) beyond individual particles. That is the limitation of computer games, just to note – their cause-effect programming is limited to the limits of the cause-effect knowledge of the programmers – a computer game character is limited in its 'senses' (the 'if' side of an 'if-then' situation), and thus responds to its environment in limited ways, based on the limits of its programming. Back to our physicist - a physicist also does not take into consideration social influences, which often do not make any sense at all (usually being superficial and fickle – exemplifying the very nature of the chaos system that life exists in). So a particle physicist can only describe how things work on an atomic level, which may have very limited, or no, bearing on what is really going on at higher levels; and it may have limited or no bearing on solving a problem at hand (such as why people kiss, for example) – to try and describe it mathematically or on a quantum physics level will most likely be fruitless, if not completely futile.

Max Tegmark: "We can deal with cosmic threats if we can just get beyond our own self-destructive stupidity."

Max Tegmark: "It doesn't fit the data." (Concerning certain skepticism's, such as over certain theories, such as the expansion of the universe and the Big Bang). For example, it is wrong to be skeptical toward the expanding universe theory after observations fit predictions (such as the 25% helium content of the universe calculated from the model of the Big Bang, and then being independently observed. Such close correlation is not coincidence).

My Observation: Science runs on skepticism, so your skeptical notions have already been covered and tested if a hypothesis had made it to a 'Theory' rating (where a theory, even as just a working model of reality, can predict things, while a hypothesis is still an untested guess).

Just to note, scientific projects are selected for funding (assuming a limited pool of funds) based on the nature and number of outstanding questions that the project is pursuing and the lack of a working model in the area; and the more pressing the questions, the more likely the funding. Also, the higher number of question marks addressed, the more likely the funding.

A Few Interesting Images:

THE FUTURE OF LIFE – EXISTENTIAL RISKS:

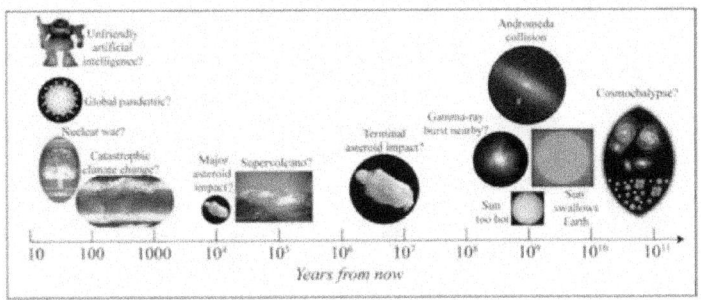

(so if man can survive his own stupidity, his chances of survival increase in the long run).

Electromagnetic Spectrum:

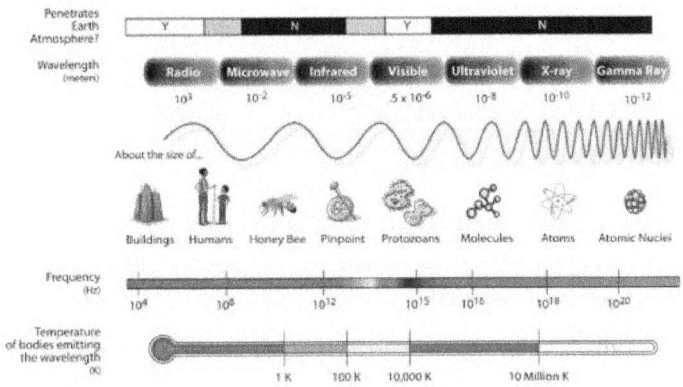

(you can see that 'radio waves' are much longer (lower frequency) and thus weaker than microwaves and visible light, and a lot weaker than ultraviolet light and x-rays.)

The Standard Model:

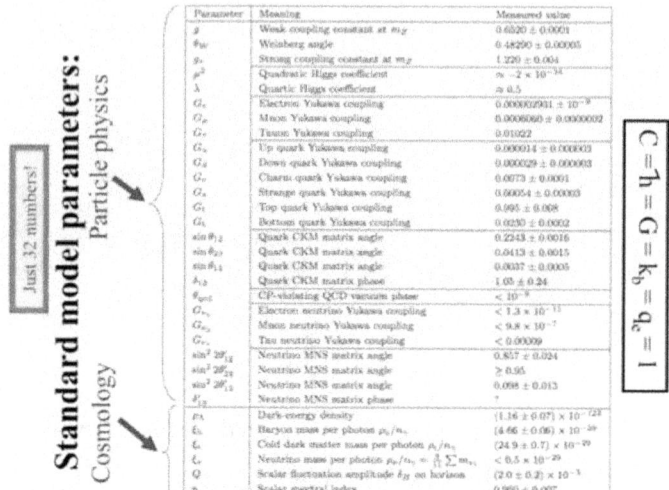

(you can see that when someone refers to the "Standard Model", they are referring to a set of mathematical constants, and not a paper cut-out of the atom or cosmos).

Critic: Max Tegmark has a new book out, entitled <u>Our Mathematical Universe</u>, which is getting a lot of attention. I've written a review of the book for the Wall Street Journal, which is <u>now available</u> (although now behind a paywall, if not a subscriber, you can try <u>here</u>). There's also an <u>old blog posting</u> here about the same ideas.

Tegmark's career is a rather unusual story, mixing reputable science with an increasingly strong taste for grandiose nonsense. In this book he indulges his inner crank, describing in detail an utterly empty vision of the "ultimate nature of reality." What's perhaps most remarkable about the book is the respectful reception it seems to be getting, see reviews here, here, here and here. The Financial Times review credits Tegmark as the "academic celebrity" behind the turn of physics to the multiverse:

As recently as the 1990s, most scientists regarded the idea of multiple universes as wild speculation too far out on the fringe to be worth serious discussion. Indeed, in 1998, Max Tegmark, then an up-and-coming young cosmologist at Princeton, received an email from a senior colleague warning him off multiverse research: "Your crackpot papers are not helping you," it said.

Needless to say, Tegmark persisted in exploring the multiverse as a window on "the ultimate nature of reality", while making sure also to work on subjects in mainstream cosmology as camouflage for his real enthusiasm. Today multiple universes are scientifically respectable, thanks to the work of Tegmark as much as anyone. Now a physics professor at Massachusetts Institute of Technology, he presents his multiverse work to the public in Our Mathematical Universe.

My Observation: If Max is like me, he will "persist", to quote you, until he is intellectually satisfied, and feels he has explored the issue thoroughly. I cannot blame him for following his enthusiasm, and it is notable that you observed and mentioned that.

Where would I take the issue of multiverses? I would leave it where it belongs – as pure speculation, or partly as postulates based on what is currently known, endeavoring to extend the knowledge into what mostly likely 'is' out there. I would invite all and everyone to forum their wildest speculations on what is beyond what we currently know, accumulating the more informed ones into one category, and classifying the others as to the wildness of their speculations – all good fodder for science fiction.

I suppose my main thrust would be to challenge theoretical physicists to explore alternate conclusions, for it is my view that, in a paper, one should give at least eight different possibilities along with the one being proposed, if only to shed light on the proposed conclusion's probability.

I think that the key component in any new theory is accurately extending what is known, or at least making sense of new data, or seeing new revelations in combinations of data from various fields. So "extending what we know" into "what is most likely beyond that" is what the whole exercise is about. If everyone agrees, then the job was well-done, though that does not validate it, since it is as yet untested and unproven – it is only a good guess. If it can be used to make accurate predictions, then yes, there must be some validity to it.

Critic: *"Tegmark's innovation is to postulate a new, even more extravagant, "Level IV" multiverse. With the string landscape, you explain any observed physical law as a random solution of the equations of M-theory (whatever they might be...). Tegmark's idea is to take the same non-explanation explanation, and apply it to explain the equations of M-theory. According to him, all mathematical structures exist, and the equations of M-theory or whatever else governs Level II are just some random mathematical structure, complicated enough to provide something for us to live in. Yes, this really is as spectacularly empty an idea as it seems. Tegmark likes to claim that it has the virtue of no free parameters."*

My Observation: The value of what Tegmark is presenting is not whether he is absolutely right or not, but in offering a new frame of mind with which to approach the unknown, adding yet one more mental tool in our observational toolbox (and we will never have enough tools with which to perceive the unknown – but that doesn't mean we can't keep adding to them).

So I like the point of view that 'local reality can be broken-down to a local set of physics equations', and it is up to us to discover the local nature of a locality (the locality being an entire universe in this case). What common sense would tell you is that the physical parameters of a 'universe', especially one that began with a Big Bang, would be determined by the size and composition of the originating singularity, so it makes perfect sense that other universes may have different physical laws, or at least different values for common laws, creating a different environment, and unique challenges for life to begin and flourish, which, from what we've seen of 'life', there is a good chance it can begin and flourish in almost any strange environment, if not any and all (if we find 'life' at the center of a star, then that would expand our view of the range of life-allowing environments, which, right now, are limited only by temperature – for we have already found microbes feasting in all kinds of chemical environments and pressures).

Critic: "In any multiverse-promoting book, one should look for the part where the author explains what their scenario implies about physics."

My Observation: The main point that any speculative notions make would be that our physics is incomplete, that our knowledge of the physical universe is not total yet. Even better, it will offer new exploratory paths and endeavors for science via new questions or challenges, or offer new frames of mind with which to pull from our toolbag of perspectives, and with which to approach existing problems and questions with.

Critic: "There's only small part of Tegmark's book that deals with the testability issue, the end of Chapter 12. His summary of Chapter 12 claims that he has shown:

'The Mathematical Universe Hypothesis is in principle testable and falsifiable.'

His claim about falsifiability seems to be based on last page of the chapter, about "The Mathematical Regularity Prediction" which is that:

'physics research will uncover further mathematical regularities in nature'.

My Observation: Mr. Tegmark does seem to be fixed on 'patterns' overlooking the role of chaos as an environment in itself, with no set patterns (which is what his Level II universe proposes – different sets of equations for different localities, which is the essence of chaos), which, in a practical sense, if right, will be critical in our physically navigating safely and effectively through such localities, using their own physics in turn to travel within and pass through them..

Summary

So as we go about exercising our will in making minute improvements in our lives (which we all like to do), we can ask ourselves, "What mathematical model describes our exercising our will?" (and the answer will be 'none', for our will is born of a chaos system (where mathematical structures do not apply, except in hindsight and in static conditions), and our will exhibits chaos itself, having adopted the advantageous qualities that such an unpredictable system offers (there is survivability value in being unpredictable), and offers to a still-developing entity that knows little about its environment and must engage it in ever-new exploratory ways. If there is one proof against a purely mathematical model of the universe, it is the chaotic nature of life itself, with infinity contributing its conundrum.

Journal 73: Large Asteroid Impact – Two Physical Effects and One Social Consideration

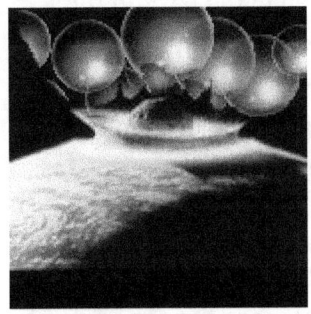

Go ahead, compare me to one of Gulliver's Laputians - people who can't sleep because they worry about the sun burning out and other long-ranges issues. I can sleep, and I've addressed the worrying with a Bell Curve model depicting reasonable amounts of resource expenditures on issues ranging from local/immediate to broad/far-ranging.

Here, I'm exploring what I think are new aspects of a large asteroid impact.

The math will have to be done to determine if the following would occur, and to what extent, but here are a few additional effects that would be caused by a large asteroid impact:

1. Massive Land Waves – This is where the land would ripple like waves, but on a massive scale. Now, at the crest of a land wave, the earth splits open. At the crest of a massive land wave, the split would be large enough to swallow entire buildings, and not just single humans, as has happened in the past. In this case we could lose entire cities, and more – all evidence of human existence in a wide area. If the asteroid were large enough, this could occur on an entire tectonic plate, which brings up tectonic plates.

2. Massive Instant Tectonic Plate Shift – We know that tectonic plates are separate, and yet are wedged together like puzzle pieces. It might happen that some effects of a large asteroid impact may be localized to that particular tectonic plate. Sudden massive movement is another matter, and could be catastrophic not only for that plate, but for adjacent plates as well.

These are two hypothetical effects that I haven't seen addressed yet; maybe I haven't searched sufficiently, but I'm presenting them here nevertheless.
Many effects have been addressed, and have had the physical math done to determine the extent of the reaction, such as volcanism, atmospheric damage, climate damage, and tsunami damage.

Social considerations, just to note, would be how to survive the impact and aftermath. Living in the atmosphere is hazardous. Living underground is hazardous, as the vibrations could shift or collapse the living space. That leaves two alternatives – leaving the planet and living underwater. Leaving the planet offers the maximum protection, provided you can avoid the path of the asteroid. Living underwater is like living in a protective gel – the water would absorb any shock, and the habitation, if it were advanced enough to be self-contained, would not rely on the atmosphere or climate, both of which could be rendered inhabitable by us, either permanently, or for tens, hundreds, thousands, or millions of years.

If underwater habitation occurred before space habitation, then the challenge of developing space habitation technology in an underwater habitat would present itself, along with the challenge of keeping in touch with events in space.

Journal 69: Theory on Dreaming

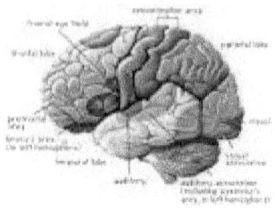

Short Introduction

Listening to a recently published scientific book on the human mind, I was mildly surprised (and a bit disappointed) to learn from one chapter that scientists do not know 'why' we dream yet, or even whether it serves a function or not (let alone 'what' function).

So I've been thinking about it, and I have a theory - still nebulous, and more creative than anything else - though it IS based on observation (if only self-observation), and it IS my aim to successfully describe reality - and not fabricate myth or an obfuscated mindset (usually out of sheer capriciousness).

True, it may not be possible to successfully describe what is really going on with dreams through sheer self-observation, reflection, and reasoning - without supportive scientific financial grants and experiments and mountains of data to interpret (either erroneously or correctly, as science goes) - though I could apply for a grant 'just to think' - but what are my chances with all the frauds out there already applying for just such grants, and without a reputation (deserved or not!), my chances are nil there... so I think on my own time and on my own self-support (one benefit being I do not feel obligated or pressured to perform (which is, strangely, conducive to performing, in this arena at least, though I would like to test the 'other side' (performing under obligation and pressure)).

My Theory on Dreaming

It has always seemed obvious to me that the nervous system plays a huge role in dreaming, but it is never mentioned. Now I've 'observed' (and which is why I am writing this now) that emotions also play a part - though they are 'emotions' only in a chemical sense - our body's natural responses (even as we sleep) to the chemical goings-on inside us and as a response to external inputs, and here is where the brain comes in.

The brain, as we sleep, as you may well know, passes in and out of REM - the 'Rapid-Eye-Movement' phase in which dreams occur (around four times a night, sleep scientists have averaged) - meaning in that state, according to my theory, the brain is 'awake' and semi-active - where it CAN sense the body's sensory outputs, and it CAN 'process' that data.

On a historical aside, this brain activity while asleep may be a primal survival mechanism that has itself survived, or has contributed to our survival (in its small way) within the frame of evolution, which is why we all have it - it must have served a critical function for millions of years, where the brain is ready to react to threats, but the body still needs rest.

It appears to me (after reasoning) that the brain, when it is 'awake' and the body is not yet awake ('repaired'), does not know 'you' (as a collection of your body and brain) are still sleeping. and yet it can detect (through the nervous system) AND process the sensory outputs that your body generates.

A practical explanation for this that the brain is ready to wake the body if a threat is 'detected' - either external or internal, which in turn is transformed into nerve impulses, which, when in 'light' stages of sleeping, the brain can detect and process, and rouse the body if needed, most likely with a jolt of fear-induced adrenalin, so 'you' as a whole can respond to reality and save yourself (and by extension, it).

The problem is in the brain's processing of the sensory inputs in a sleep state (and 'dreaming' can be referred to as a 'problem' when seen through the lens of my theory, though not a critical one, since the brain is usually wrong).

What is happening is that the brain is detecting sensory inputs, from external factors (such as noises and temperature changes), and internal factors such as feelings (based on what state your body is in - and this includes ailments, how fit the body is, and what is being digested), and emotions (a combination of the particular body feelings and memories).

Let's explore the role of emotions and memories. It appears that dreams are indeed a combination of your body's sensory data (responding to external and internal sources) and your particular memories. So just what is your brain doing in that situation?

Here is what your brain is doing - it detects these sensory inputs (now that it is 'awake' in its own way), and, in the absence of visual references (since your eyes are closed) it tries to 'make sense' of the data - it 'reasons' that 'something' 'must' be happening, and what does it do? It attempts to create that 'something' itself - it, in effect, tries to fill-in the 'explanation void' (which it does out of habit - it usually receives visual explanations when awake).

So let's see where we are - the body continuously sends the brain sensory inputs, even when we are sleeping, and the brain, awakened from its own sleep (our 'deep sleep' stage), begins detecting sensory inputs from the body, and it can't help but try to process them - that is what it does. In the absence of visual references, it tries to fill-in the data void - generating what, from your experiences, it thinks MUST be happening 'out there' - and, curiously, it 'sends' this data to another part of itself - the part that processes vision.

So the question was, where does the brain 'get' this substitute visual data, and the answer is from memory, and not just any memory - but those memories that were obtained when your body was in a similar sensory state (caused by a similar combination of external and internal factors).

Why are dreams so weird then? Your memory - it is well known that memory is quite faulty, and all the various fragments that your brain pieces together based on your body's (continuously-changing) state can be quite surreal and abstract, not to mention entertaining, and even horrifying.

So there your brain is, receiving sensory inputs, from the sleeping yet still-alive body, via the nervous system, and the brain, out of sheer habit, expects to receive associated visual references. Well, during sleep, it ISN'T receiving any, and this is not right - this perplexes the brain, or perhaps the brain is alarmed that there isn't any - which places your (and its) survival at a disadvantage, so it tries its best to supply what must be 'out there in reality, which, now that there are visuals, primes the body for actual response (which is why we 'jump' and twitch (even shout and scream) during REM sleep).

Conclusion

Well, there you have it - I've covered a few points, and the theory seems to be holding-up - and that is the key factor with theories - whether they hold-up to new considerations, and any and all counter questions, considerations, and tests.

Now I am getting weary again (it is the middle of the night after all), and my body is telling my brain that it isn't sufficiently rested... or perhaps I have worn my brain out, and IT is telling my body that it needs new rest - and imagine that - a reverse data path, where the body detects and processes sensory signals from the brain (outside of motor commands - meaning the needs of the brain itself) - perhaps this is where hormones come into play...

Where Does One Go With a Theory?

So where do I go from here with this theory? I can do a Newton and just let it sit for three decades until a Halley comes along due to a wager and spurs me to publish it; or I can do a Darwin and ponder it for two decades, to see if it stands up to new revelations that come with time, before I take it seriously, or until a Wallace comes along with the same theory (only derived at through sudden insight, and not through two decades of work and testing), and startles me into publishing.

I cannot approach scientists with this - they are caught-up in their own pet theories, and will, defensively, scoff at anything new. If there are large sums of financial grants at stake, then things get much worse - in the words of one scientist, "a lot of bad science gets thrown around - hasty conclusions that do not stand-up to scrutiny; and science get very ugly when playing at a certain level".

Since I do not feel like dealing with that monster, I'll just post it here, see where current studies are, and ponder the matter further, at leisure, Wallace or no Wallace (since I can!).

I will probably, as part of the process, casually bounce the theory off of people when the occasions arise, to test if they are inclined to agree, or to learn their experiences and thoughts on the matter - but the problem with 'casual' is that most people are glib in those situations - they like to hear themselves talk, whether accurate thoughts are behind them or not - though there is still value to even this - for whatever the points are, they can be used as challenges to one's theory - and if the theory holds-up against counterproposals, it is still alive, if not, it is dead, and should be allowed to die (and given a proper burial).

A problem in science is that people tend to become attached to things, their 'theories' in this case, and they become unwilling to 'let go of them' (discard them) (let them die) in the face of more reasonable counter theories, or when their theory breaks-down in the face of something that was not accounted for.

In the case of dreams, whatever people glibly spout-off WILL be purely speculative, even if they are referring to science, since science itself is still unsure.

Agonizingly (for me at least), what people usually glibly spout-off (and I'm speaking from experience) is derived from current (or past) fantasies, occults, and/or common myths, rather than from independent objective observation (self-observation included) and 'insight' (derived from reasoning, whether deductive or inductive).

It may be wise that they do not trust themselves, yet, I would much rather have objective observation and independent thinking than be served servings of current fashions in fantasy, occult, and myth - though I do admit that they can be fun, and they can be conducive to social success; but these are not my goals here...!

Journal 65: Current Views of Life and the Universe

Reference Video: "Are We Real" (the title of the video post), or "What We Still Don't Know" (the title of the actual video)
https://www.youtube.com/watch?v=fRzPM3FgF9I

Overview

The video addresses the current mentalities in cosmology, physics, mathematics, and philosophy, concerning the nature of the universe (and of the life in it), and current leading speculations about what is yet beyond us.

I've presented all the concepts covered in the video, and I've offered any thoughts I had that exposed them as wrong, or set them right; or, if the concept was OK, that elaborated and delved further.

Interesting Video Statement:

Four fundamental questions in cosmology:
Was there a beginning?
Are we alone?
What is the future of the cosmos?

What is the nature of reality?

Interesting Video Statement:
As we progress, more questions arise, or, more specifically, more detailed questions arise within that category, such as, "How widespread is intelligence in the cosmos?"

Interesting Video Statement:
"We are the most complex things we know of in our universe, particularly our brains."

Interesting Video Statement:
Of all creatures, we are the most special.
My Thoughts:
I would add, 'currently' – in that we have just entered our 'Brain Age' while other creatures have not entered theirs yet.

Interesting Video Statement:
Religions gave imagined answers for reality, science discovered the truth.
My Thoughts:
Who is at fault? Religion, for overstepping its bounds – it got into playing the 'is' game, and was exposed as fraudulent numerous times.

Interesting Video Statement:
Responding to the religious belief that "we were no accident waiting to happen": As scientific knowledge

grew, 'man' received a series of demotions to his ego, where, at his ego's height, we were the center of the universe, the center of its concern, the center of 'God's' concern.

My Thoughts:
As the current supreme being on earth, we still are in a way – though only locally and currently.

Interesting Video Statement:
Science discovered that the universe, including life, is guided by the laws of nature, the most fundamental being those under the study of physics – laws of nature such as the speed of light, the force of gravity, and the charge carried by an electron, all with precise values, and all determined by the nature of our particular Big Bang.

Interesting Video Statement:
The "Life-Survival-Death" Board Game (the "Game of Life" or "The Life Game" – John H. Conway (mathematician)): (an 'artificial' biology) demonstrates that complexity develops out of simple rules, which we tinker with. The three simple rules of the board game of 'Life' are Birth, Death, and Survival. You begin with a grid. You begin with a random number of playing pieces ('counters'). Any empty square surrounded by three other counter gives 'Birth', and a new counter is added to the board. Any counter with no neighbors would 'Die' (of isolation) and is removed from the board. Any counter with too many neighbors (four or

five) would 'Die' (of lack of resources) and is removed from the board. Therefore any counter with just two or three neighbors will 'Survive'. Unpredictable and complex patterns 'evolve' – the board seems to produce 'creatures' (repeating complex patterns) from nowhere – creatures that 'move' (recurring sequences of patterns that move across the board) (think single-cell organisms), creatures that fire out smaller creatures (think 'white blood cells), patterns that display pump-like actions (think 'beating hearts'), and creatures that spew out an endless chain of offspring (think 'replicating DNA'), all behaving like a mini universe of life, and all existing on a random foundation.

My Thoughts:
I'd like to play that game...

Interesting Video Statement:
"Therefore there is no design in life whatsoever."

My Thoughts:
But it is still possible that 'someone' designed the basic rules, the foundation upon which the randomness exists, as in the game – someone had to create the counters and the board, and set up the rules.

Interesting Video Statement:
"We are the products of atoms and time."

My Thoughts:
I would add many other factors (though they may be merely sub-factors) such as forces and manipulated

probabilities (though the force behind the manipulation is randomness).

Interesting Video Statement:
"'Life' gives meaning to the universe."
My Thoughts:
I would add, 'for itself' – for the universe is not anthropic, and physical matter and energy do not need meaning to exist.

Interesting Video Statement:
There is the notion that the universe was fine-tuned so that we could exist; and the most exacting currently known parameter is the cosmological constant, which is 'fine-tuned' to a degree of 1 in 10-to-the-120-decimal-places, and where which, if it were off by just one unit, the universe would be so vastly different that we could not have evolved as we are.
My Thoughts:
That the universe is fine-tuned for us looks at the issue from the wrong perspective, since we exist randomly, and are merely one of many random results of such tuning. The question should be, "Is our universe the only one that life can exist in, or can life spring forth in any universe? Meaning changing parameters of our 'universe' may be bad for the continuance of our particular life forms, but not life in general – where life would spring forth within any set of natural laws and forces, in any universe, no matter what 'form' it took (which would be a form of necessity), or on what scale.

So we are the random results of the fine-tuning of laws and forces to the existing exacting parameters.

Interesting Video Statement:
On the notion that "We are lucky to be alive."
My Thoughts:
We are lucky individually, but not 'lucky' in regard to life in general. Our present forms are merely the result of our environment (the laws of our particular 'universe', and our physical surroundings). In other words, only the values of the laws of nature (and the surroundings) determine the form of life, and generally not whether life will spring forth or not (though in extreme universe it will not).

Interesting Video Statement:
Why are the values of the laws of nature set as they are (for example the value of the electric charge of an electron or the speed of light) – and the only answer is that they were based on the nature of the particular Big Bang that spawned that particular 'universe' (as we've been calling its resultant matter and the space it occupies – my note). Different universes may have different values of particular forces, or different forces altogether.

Interesting Video Statement:
Scientific 'Uncertainties' are that which are still beyond experiment or test.

Interesting Video Statement:
It may be possible that the reason the universe is the way it is has to do with our own existence (intelligent life), which create this exact universe so we could exist.

My Thoughts:
My answer is that, in the scheme of things (given evolution of all forms of life), this is preposterous – since 'we' are but a small part of the whole of life, and a mere footnote the course of evolution.

In regard to our universe being created, it could be that our universe is so complex because it is a Model-T universe (referring to Henry Ford's first production cars - an American's favorite reference for something primitive, inefficient, ugly, slow, and obsolete) – a primitive attempt before the production process became streamlined (and thus more simplified, more elegant, and more efficient) (though by future standards such streamlining itself a Model-T!).

Interesting Video Statement:
Our present views and knowledge will, as history has shown, become very restrictive in the light of future discoveries.

Interesting Video Statement:
We have scientific explanations of the universe.

My Thoughts:
There are 'scientific' explanations, and then there are explanations based on logical analysis (resulting in 'theories'). The former are accepted through actual testing and experiment, while the latter (logic) are accepted when they stand up to all questions (adequately) and stand up to all challenges from alternate reasoning (adequately), and I say 'adequately' since the explanation of reality can become infinitely complex, though most of that does not apply to us in any 'practical' (adequate) sense, so 'adequate' means that which matters in an immediate (practical) sense.

Interesting Video Statement:
There may be different forms of life in different 'universes' (all the matter under the influence of one particular 'Big Bang').

My Thoughts:
The forms of life in alternate universes (those with different sets and values of natural laws) can be calculated (based on those sets and values), including the ultimate allowable level of complexity for that life (for example, on the early magma earth, life could only reach a certain stage of complexity, such as single-celled organisms - the environment being so harsh). Can we currently do the calculations? No.

Interesting Video Statement:
"By the time the sun expands and vaporizes all

remaining life on earth (in six billion years or so), the supreme form of life on earth may not be 'human' in form, but may be at least as different from humans as humans are from the single-celled organisms that existed three and a half billion years ago.

My Thoughts:
I'm willing to bet that a lot of new technology (rather than new biology) will be incorporated into such beings by the beings themselves.

On The Form of Future Intelligent Life:
As biological evolution goes, our sense may improve, and we may develop new senses that would become a part of future intelligent life.

Technology offers scenarios beyond purely biological intelligent life, where future beings have incorporated enhancing technologies into themselves – for example those that enhance our senses and physical processes (such as strength, speed, and precision), and our mental processes (such as performing mathematical calculations). You can see that we are already technologically enhanced beings, so the issue becomes the level and magnitude of the enhancements that the technology brings (successful evolution being measured by the probability to survive and perpetuate into the future, which makes the assumption that increases in senses and physical/mental features and abilities will aid in survival and perpetuation).

Interesting Video Statement:

Those of Nick Bostrom (philosopher).

My Thoughts:

Note to self: Interesting modern philosopher. See what other thoughts he has had.

Interesting Video Statement:

One may one day make backup copies of oneself, until one was were made, for example, of purely silicon. The technology would begin by replacing one nerve cell with a silicon-based replacement, for example.

My Thoughts:

'You' at present are defined by your 'limitations' (the limits of what you know and can do), which brings up the question, "Why would you want to be you as you are at present, you being so limited? If you could recreate yourself, why not recreate yourself with enhanced abilities and knowledge, or with all knowledge and all abilities? God believers would find this offensive – how dare we desire to become Gods! (it is why Adam and Eve were expelled from the Garden of Eden, after all); but what is 'God' but man's own image of the ideal man? (or at least a 'super' man, given the less-than-ideal characteristics of the ancient Greek gods)

Interesting Video Statement:

"It doesn't matter if we are made of silicon or carbon, it is what is in our heads and our feelings that matter."

My Thoughts:

If we eventually make ourselves out of pure silicon, that removes the necessity for 'sex', since we can simply 'build' babies. So a substitution for all the benefits that sex gives us (and it is a good bet that many are as yet undiscovered) would be required (while hopefully leaving out all the detriments, such as all the pain and burdens being pregnant causes women, not to mention their possible death in an excruciatingly painful childbirth).

Interesting Video Statement:

What can a future super-intelligence do? We may never know.

My Thoughts:

We cannot know specifically, but we can formulate general rules – it will be limited to dealing with matter and energy, for example, on some scale between nothing and everything, and within certain boundaries within infinity and eternity. (so there, future super-intelligent beings, gotcha…!)

Interesting Video Statement:

Since computers can simulate, could we merely be a simulation? or, "Are We Real?" and yes, we are real for all of our practical purposes.

My Thoughts:
This is a speculative possibility, along with an infinite number of others, and it remains to be tested.

Such a computer could go about creating a sim in two ways – set up the basic laws of the universe and 'let it go' on its own, randomly and by chance; or try to control specific outcomes (such as our present forms), which seems far less likely or possible, given the complexities involved, and which through chance (requiring much less effort) can already come about naturally. Such a specific-form controlling computer, however, would have to have the ability to manipulate energy and matter precisely and in three dimensions, which they already do, but as of this date only as visual projections.

If we are simulations, are we 'real'? The answer is yes – to us, but perhaps not to the builders of the simulator. It does affect how far we can go, or at least the particular challenges we face getting somewhere or achieving certain goals (such as overcoming death and ignorance, and perpetually pursuing new knowledge and dealing with newly-perceived threats – in the sim case it would be having someone close the program or the computer losing power for some reason (intentionally or not).

Another thought is, even if we are simulated, we can also create simulations, so you might end up with an infinite number of simulations within simulations, which begs the question, if someone closes their simulation program, will it affect all lower-order simulations? (you would think 'yes') – which means just turning off such a computer will affect the infinity that it has created, and since we know infinity does not 'exist' (it has no bounds – and 'things' must have bounds), that is a paradox.

Another scenario that might occur is that a simulation might gain control over its simulator (made easier if the simulator is a simulation itself).

Thinking about a 'way' to create a 'simulated' universe, all you have to do is throw-in a lot of particles, either in a closed system, or in close proximity (as was our case, though our 'universe' is dispersing), and then let them intermix and interact, and life will spring forth.

Another question is relative size (and the associated relative time-speed) - what would be the possible ranges of relative sizes of the simulated universe to the simulator? You would thing that simulator would find it harder to create a simulated universe larger than itself, and easier to create on smaller; then the question shifts to 'how smaller', and are there quantum size limits in order to prevent detrimental interactions between simulator and the simulated (assuming a simulated universe must be made out of matter or energy, which seems to be a reasonable assumption!).

One fallacy in the physically-oriented simulation approach, such as creating the simulated entities as far down as individual neurons, is that this is the wrong approach entirely - since such entities are the result of events and outcomes, all random and chaotic, and all emanating from a 'beginning', so it is more practical and far easier to just start a simulation at the beginning, with a load of atoms, for example, and just throwing them together (as was most likely the case in our existence). If you wish to recreate individual physical objects such as brain cells, you can only do so for one static point in time - after that, chaotic events and their unpredictable outcomes will take over (if your simulation has movement and flow, and is not just a 'still photograph'; so if you 'build' a simulation with the physical approach, you can only 'build' a starting point.

After delving further into the topic (see https://www.youtube.com/watch?feature=player_detailpage&v=nnl6nY8YKHs ("The Simulation Argument"):

The notion was propounded that you could 'rewind' the simulation, and this supposes that all events and physical outcomes are 'stored' - where you could 'begin' the simulated universe at any point, and run it from there again (or play it backwards, just to note that, or at different speeds, or in combination with other universes, as if you spliced two program flows together at their outputs). In recording your simulation, you have two options: One, where the 'recording process' only records, it does not 'control' outcomes; and Two, where you've constructed all the outcomes in advance for the entire time span of that simulated universe, and all that remains to be done is to choose a point in time, and press the 'play' button, which presents all kinds of problems for the simulated - popping into and out of existence, for example, and a loss of continuum (which we do not experience).

Interesting Video Statement:
"Logic has taken science back to where it started – could our super-intelligence be the God we have always imagined?"
My Thoughts:
Like I said, 'God' is man's creation of what an ultimate man would be like – creating things within the bounds of 'good' (though here it is evident that even the Christian God screwed-up).

Interesting Video Statements:
"From simplicity (a few natural laws) comes complexity." "What are the 'philosophical implications' of a philosophy." "We are pursuing fundamental questions about existence and the nature of world."

Interesting Video Statement:
"All reasoning and evidence ultimately returning to an all-powerful, all-knowing super-intelligent being that has created everything."

My Thoughts:
If so, everything but itself, which must have then been created by pure chance and randomness driven by a few simple laws of nature and their resulting forces - meaning 'nature' came first; so the question is what laws existed in that nature to allow such a being to evolve so far, and do we have to emulate it to achieve like status?

Interesting Video Statement:
"Science should not dismiss ideas that are 'weird'."
"Anything but 'weird' would be a big letdown."

Interesting Video Statement:
"It is amazing that we have made so much sense of the external world."

Interesting Video Statement:
"The challenge for the next century is the question, "What do we still not know?""

My Thoughts:
I would argue that it is rather, "Why bother?" which precedes asking all other questions, unless the question is a natural one, then it precedes dealing with the answer, unless the answer is naturally dealt with!
And even though we do not know what is 'out there'

(beyond our limits of perception and current logical reasoning), we can identify the best paths for the immediate future to investigate (until further knowledge indicates other, more logical (wiser) (more relevant) paths).

Journal 63: Underwater Human Habitation

I've covered my notions (original and innovative, as usual, if not just odd) on this topic in detail before - I just can't find it (I think it was buried inside something else).

I want to give this its own Journal, so I can quickly return to it to add further thoughts and details.

My initial impetus for this notion was revitalizing the heavy industry of Detroit (of all things!). The notion was due to Detroit's being in the proximity of the Great Lakes (and in need of revitalization). An additional item also spurred this - reading about the old telegraph cables that were lain across the Atlantic Ocean in the 1800's, where, when a segment was pulled up for repair, they found all sorts of life on it, which struck me as a way of generating food for such an underwater human habitat.

I recently made a little 'funding proposal to Congress' speech that I wanted to add to the main body, but so far I cannot find the piece that I wrote it in!

So the speech here will have to do for the time being (which itself was buried in a poem and not easy to find)... to pick up in the middle of it...

"We need this not only as an engineering exercise,
which is worthwhile in itself,
and which would result in many collateral engineering
benefits
(as such endeavors always do – just look at NASA);
and not just to create widespread employment in its
implementation,
employment being a popular political prize for any
competitive political parties such as yours,
and to the narcissistic, egotistical, fame-seeking,
power-grabbing career politicians in it, like yourselves,
but also for a more pressing and practical issue –
to create a potential aqueous buffer to life-
annihilating, cataclysmic events
such as the scorching, choking, debris-filled aftermaths
of large asteroid impacts
or super volcanoes or mega-earthquakes or super-cell
weather systems
or extreme climate change,
(though that is not to say that such life-annihilating
calamities do not have benefits -
for they would solve the problems of many people
struggling in life),
calamities which, if we existed solely on land, would be
catastrophic to our species.

So, in effect, it enables us to better (if incrementally) survive planetary change and instability of ever-increasing severity.

All of this is within our general pursuit of leveling-up within our Brain Age –
where we are able to withstand more of what the earth,
and then the universe, throws against us and the life around us –
which addresses our own survival as a reproducing-capable species,
and, beyond that, us in the role of life's guardians and caretakers
– at least land-based;
where we have reached that 'next level' of the 'Brain Age' - that of being guardians and caretakers,
and where we should not shrink from those duties –
for life is still in a trial-and-error stage of evolution,
and we do not know which will succeed, meaning we need to recognize and value diversity
while we continue to explore, experiment, hazard, and face the challenges and problems
(at least those we are aware of) of life in this universe;
and all this is not even mentioning the uplifting of human (and less, national) spirits
that such endeavors and adventures bring (as well as new insights),
uplifting because each individual has input into the

endeavors

that will expand our adaptive abilities and existence options as a species,

and well as being a vehicle of social and mental progress and health.

In summary, we need more people who can create worthwhile enterprises

that create useful mass jobs that people can center their lives and families around,

and this government has the opportunity to fund it, right here, right now;

and if not this government, then I will find another government –

though it would have to be a state that values individual freedom and input,

which rules out tyrannies of any form and their inherent repressions and weaknesses...

So, in summary of my summary, even though what we design and build will be primitive by tomorrow's standards -

meaning large, bulky, slow, inefficient, and ugly, we need to build 'something', anything, for future generations to improve upon."

Further Details

Here is why I posted this topic as its own Journal - I had additional thoughts on the matter:

An Artificial Magnetic Shield

One purpose for engineering an underwater human habitat is as an incremental addition to our ability to survive cosmic calamities such as large asteroid impacts, which would destroy the paper-thin atmosphere that surrounds our planet that we currently depend on,

and as a buffer for the resulting earthquakes, the photosynthesis-killing post-calamity millennia-long 'night', and the resulting extreme weather systems and shifts. Such an ability (living underwater) would free us of that dependency, and aid us in our eventual migration from earth, giving us the ability to exist in unfriendly environments.

A new idea occurred to me, however. We also rely on the earth's natural planetary magnetosphere to shield us from deadly cosmic radiation (mainly from our own sun). Being able to create an artificial magnetic shield would free us of this dependency, too (as we gradually free ourselves from all of our dependencies on earth).

Such things are most cost-effectively worked in at the planning stage, though usually they are not, not being initially important, and better solutions and applicable technologies often appearing later.

Precursor

I found what I had before - it was buried in "The 400 Year Old Man" (fitting - since it would take years):

The details of the deep-water industry occurred to him first – forming from nebulous, romantic visions of underwater cities – for what purpose he would build them he did not know at first, it was more out of a frontier spirit – that, "There the water was, and here we are, and it was just something that we should be doing." Then more practical reasons slowly formulated, giving the city an immediate purpose –

and the first purpose that occurred to him was food production – farming the ocean floor, or Great Lakes floor – with either fish or with edible, nutritious plants, or both... theme parks would follow later...

There were also all the other traditionally land-based opportunities that were equally available underwater – the equivalent of railroads and mining operations coming to mind... and, of course, at the forefront, or at least parallel with all other engineering endeavors, scientific exploration and discovery... and yet, in the spirit of his eco-conscious age, he was mindful of the existing eco-systems – the days of the Earth-raping, society-destroying robber barons were long over... he would spend time thinking of how to deal with the ass-end output of his enterprises; perhaps government would help belay the costs here, if the front-end results were worth it...

One enterprise he wanted to pursue was 'deep ocean cable fishing' – he had heard about the Transatlantic cable, and when they pulled a section of it up for maintenance, there were all kinds of deep-sea crustaceans on it – "Edible," he thought, and he imagined a perpetual operation of laying a cable, the letting the cable 'collect food' - meaning allowing the crustaceans to attach themselves to the cable and netted extensions, then pulling the cable up and harvesting the food along its length, then laying it down again, in a kind of perpetual circular arrangement –

The cable 'circle' would be the ideal length that, by the time the 'circle' was completed, new food would have attached itself to the beginning of the cable again, which would be ready for harvesting; or, less eco-stressful when needed, seeding them with eggs...

So there would be the huge main cable laying/retrieving/harvesting ship, then service and support ships, and then a myriad of world-distribution ships.

It would be a fine operation, helping to feed the world, and in keeping, if not more enterprising than, the ideals that Lyndon Johnson propounded in his "Great Society".

The Great Lakes were not as deep and the deep ocean, and the fresh water offered different opportunities, and they offered a more hospitable environment for developing underwater endeavors, especially the first underwater cities, which, by future standards, would be the Model T's of history – as sleek and fast and attractive and efficient as they would be in the beginning, no doubt they would become clunky, slow, inefficient, and ugly in comparison to future designs, so a less harsh environment would be best suited for endeavors still in their infancy.

Journal 62: Atomic Model - A "Stationary Particle" Model of the Atom

Introduction

The current standard model of an atom has protons and neutrons clumped together in a nucleus, each particle being made of quarks; with electrons orbiting in probability orbitals about this nucleus, with fuzzy 'quantum leaps' being made between a limited number of possible orbitals; meanwhile the electron is kept from falling into the nucleus by a weak nuclear force.

I'm sorry, but there are just too many absurdities with this model...!

One is the nucleus - it is said that the neutrons in it make it possible for the protons in it to clump together. That is preposterous. Another is the notion that the electrons are in motion and in orbit, and that there are but a few specific distances for the orbits, and that the electrons make quantum leaps between these few-allowed orbits - disappearing from one orbit, into nothingness, and then reappearing in another one of the few specific allowed orbit (the 'quantum leap'), and to attempt to explain why the electrons do not fall into the nucleus, a 'weak nuclear force' was fabricated - all of which are equally preposterous explanations of what has been observed, spurring me to come up with a more reasonable model of the atom.

So, in a flash of insight, I envisioned a "Stationary Particle Model", which is far more likely; or, in an alternate title, a "Layered Tetris-sphere Atom", which offers better rational explanations of the various experimental observations that the previous orbital model was based on.

Analogy: If you've ever played the game Tetris - the atom is like that, with pieces falling into place (guided by fields in reality, and in a chaos environment) in a spherical form (so there IS a Tetris game like that - "Tetris-sphere")., only we have a sphere of protons and electrons (and proton-electron-created 'neutrons').

My Stationary Particle Model of the Atom

My "Stationary Particles" Atomic Model

○ You begin with a 'nucleus', negatively-charged, at least at its outer layer, for there are an infinite number of ever-smaller suprises within it. Why negative? That is the quantum layer that our universe happens to be at.

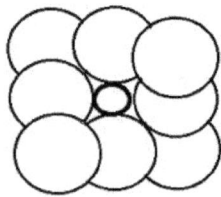

The next layer consists of protons, all stuck to the nucleus - the attractive nucleus-proton force overcomes the repelling proton-proton forces to keep the whole together. In larger elements, there are several layers of protons.

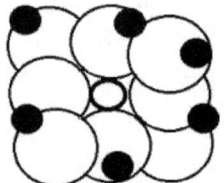

Electrons bond strongly with the outer protons, which, when each are bound together, form a neutron, creating a neutral layer between the inner protons and the outer, weakly-bound "free" electrons.

Stationary Particle Atomic Model
Simplified Schematic - Levels Valences vs. Orbital Valences,
and Levels vs. Quantum Leaps
Neutron Level 1   Neutron
                                                                Level 2

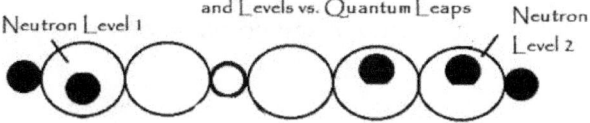

Outer electrons try to bond with inner protons, but the neutrons
are in the way, so the links are weak.

The layers of neutrons create levels, which the electrons jump between
when they are excited and de-excited, jostling about and filling the
available niches. If they 'fall' into an inner layer, they give off light energy
of that particular wavelength. These levels were mistaken as 'orbits', and
the electrons moving from layer to layer mistaken as 'quantum leaps'.

How Confident Am I of My Model?

Well, consider this: My atomic model was born of
incredulity and derived from a few flashes of insight
and few subsequent considerations, all in less than one
day, with no mathematical verifications; while the
former atomic model was derived from the greatest
minds in history, much over a century and a half of
atomic, electrical, and thermodynamic theory and
experimentation, which is now accepted by nearly all
physicists, and has already been put to use by
engineers and the military.

So, considering that, I'll give my model, oh, about
0.01% chance of being "the" model - after all, think of
the odds...!

I could look at it as creatively answering that 0.01% chance that the current atomic model is filled with absurdities and is utterly, and preposterously, wrong.

Particle 'Clouds'

Consider the 'clouds of probability' that atomic physicists see - which denote 'up' quarks, 'down' quarks, and other quarks - where it could be that they are stationary until the experimenters try to 'see' them, bombarding them with energy, and CAUSING them to bounce around in probability patterns, as Heisenberg said would happen when 'trying to see such a particle'.

In other words, the clouds that scientists 'see' are merely the 'dust' that they are 'kicking-up'. As soon as they look away (leave the system alone), the 'dust' settles again - meaning the atoms return to its stationary form - but for the disturbances in nature - it is more likely that an atom will only be absolutely stationary at absolute zero temperature.

Applying the Math

Now this model is an example of the concept coming first, then the hard calculations being done to see if it is possible, then comparing the resulting quantities (of field strengths in this case) with existing measurements, to see if they match. This has been done many times in science - a theory is proposed, and then found to be true via quantified observations and mathematical representations.

If such model calculations coincide with existing knowledge and stands up to all questions, then the model is useful; if the calculations do not measure up, then the model is one of two things: 1. completely wrong, where it is just not possible in the physically world in any variation; 2. is on the right track, but more work needs to be done.

I have no inclination to do the math myself - why, that would be selfish! I suspect it would be a lot of fun for a mathematician (somewhat less for a physicist). The mathematician's calculations would be physics-based - based on sizes, masses, field strengths, and available positions to see if one or more configurations are even possible.

Ions and Isotopes

Ions and isotopes are possible with this model.

How Molecules Would Be Made From These Atoms

Consider the currently accepted fact that, on the atomic level, there is mostly open space between atoms. So how can an atom with stationary particles 'cling' to other atoms to make molecules? Consider water, for example - with one hydrogen atom and two oxygen atoms - how would they bind to form a molecule, and how would that molecule combine with other molecules to form substances?

The "Open Space"

From our time perspective, atoms move about blindingly fast, and from our space perspective, they seem very close together; but then take the atom's perspective - to them, interacting with other atoms is just like how our stars interact with one another to us - they are extremely distant, with an unimaginable amount of empty space between them, and yet they are 'bound' together in a relatively slow (to our perspective) dance. So too with atoms - to them, they are moving eon-slow among one another, with fields that are, to them, so weak as to be undetectable (as if we tried to detect the gravitational strength between two of our stars). So these atoms, through a very weak and leisurely dance, form elements, which in turn, and in the same slow dance and weak attraction (for them), form molecules, which, in the same slow and weakly-attracted dance, form substances.

You can see how substances would be the easiest to separate, then molecules, then elements, then atoms, then subatomic particles, ad infinitum, until you're destroying the next universe within (assuming infinite space inward (just as in outward) and an infinite number of quantum-size-separated universes within at any given point).

The Nucleus

The nucleus is now filled with mystery. Physicists have only delved one level down - to the quarks, or what appear to be quarks...

On Quarks

There are two possibilities here - 1. That yes, they are actual particles, and 2. They are fragments of particles chipped off during collisions.

To test this, you would keep an eye out for oddly-sized particles that have picked-up the fragments, and others that have lost them; otherwise, if they are all uniform in size and mass, then yes, the must be actual particles rather than randomly sized particle fragments. The 'fragment' scenario would explain all the seemingly different subatomic particles - where they are merely different sized fragments.

The Interactive Fields of Particles

Particles would not interact if they did not have fields. Their electromagnetic fields have already been put to use in cathode ray tubes (which include early televisions), but the strong and weak nuclear forces may merely be ad hoc mental patches to make sense of the former quantum-leaping electrons 'orbiting' around (or kicked-up, like dust) a proton/neutron-nucleus atomic model.

On the Value of Correct Understanding - Explanations, Predictions, and Uses

A correct model will explain all that is already known, and that which is not known yet, and all the anomalies that are found in any system.

One practical aspect of a correct model is that it can be used to make predictions; another is to be able to put that 'something' to more, and 'better' (more specific and more efficient), uses.

On Instantly Rendering Previous Physicists Instantly Irrelevant

It has happened, in Geology - where the theory of Plate Tectonics turned the geological world on its head, and all the textbooks had to be rewritten overnight.

Here, if my model is correct, then a lot of work that physicists have done over the past hundred years will be rendered instantly irrelevant - like fabricated the weak nuclear force and 'quantum leaps' to explain away the improbable existence of 'orbiting' electrons.

Do you think those physicists will be happy about that? Well, considering that we've taken a step closer to reality, they should be - being scientists first; but they are humans first, and they will probably be embarrassed, then they will become defensive and angry at me and my model, and at first dismiss it out of hand, then deny it on some fabricated reason or another (and I'm speaking from a lot of historical anecdotes toward other 'new theories' - from Darwinism to the Big Bang), then someone will try to abscond it...!

On How to Present (Publish) Such a 'New' Scientific Model

I could write a dry scientific paper; but being a creative writer with a lot of spirit, I could present it in a Science Fiction novel (which most scientists would advise me to do on this matter - believing I'm wrong); or, having hung around the poetic world, I could add poetic romance to the story, even high adventure - and why not? Just think of it - a revolutionary scientific advance presented in a poetic sci-fi romance/adventure novel (and I'd have to add comedy - I can' resist that)... it would be a 'first' - though not a first in new science ideas being presented through Science Fiction - Arthur C. Clarke's geostationary communications satellites immediately come to mind; and who knows how close we are to Jurassic Park? (That is one of the benefits of Science Fiction (and any novel) - they present problems that we haven't faced yet (and some even offer solutions)...

Nobel Prize Daydreaming

Even though this model may be completely wrong, such daydreaming is inevitable, so allow me to show you what it is like (for one has to get it out of one's system before moving on!)... Now, if I am informed that I 'won' one, then I would have to insist that I receive a 'Light' Nobel Prize, since I did not do the 'heavy' math involved to verify its possibility and accuracy, nor propose, develop, or conduct the required experimentation. I would magnanimously insist the mathematicians and experimental physicists who did do all the heavy work get the 'Heavy' portion of this particular Nobel Prize. The prize money would be split 10-45-45... no, 20-40-40... no, 30-35-35...!

Spooky Action

On an unrelated related note, the principle of Newton's Cradle may offer a possible answer to Quantum Mechanic's "Spooky Action".

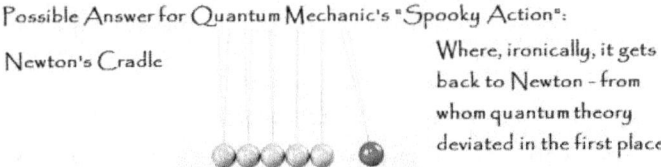

Possible Answer for Quantum Mechanic's "Spooky Action":

Newton's Cradle

Where, ironically, it gets back to Newton - from whom quantum theory deviated in the first place.

Where the seeming impossible connection between two separated particles that have interacted and seem to remain connected through a mysterious 'string' and still react to one another is just Newton's conservation of momentum and energy in action, transmitted via intervening atoms - not necessarily physically interacting - fields will do.

The obvious 'test' would be to perform a test in the absence of intervening atoms (which a vacuum would ostensibly give you - and I say 'ostensibly' because we may not know what is still there, given things like anti-particles and dark matter)...

This explanation is weak, for it is more likely that wrong assumptions are being drawn from observations.

Divergent Postscript

Those Fanciful "Micro-Universes" and This Model

Now, in the current standard atomic model, inside a hadron such as a proton, it is said that there are massive numbers of moving and spinning (making them spherical in nature) gluons which keep together a high number of moving and spinning (similarly spherical in nature as a result) virtual quarks and antiquarks at a 1:1 ratio, with the odd few remaining determining the property of the particle via their movement and spin. Further, we are told that such protons and neutrons clump together to form an atom's nucleus, so here is the challenge – envision these hadron 'particles' (protons and neutrons) consisting of their many gluons and quarks and antiquarks, WITHOUT a shell, without an OUTER MEMBRANE, so to speak, (such as a biological cell has) to physically keep all the little quarks and gluons inside, and what do you have? A lot of little quarks and gluons acting as a unit (creating the hadron particle, such as a proton) without an outer protective membrane.

Now try to ponder what could possible keep all the quarks and gluons of one particle separated from (and from intermixing with) all the quarks and gluons of the surrounding adjacent particles, particles themselves which, remember, have no physical 'membrane' (like a biological cell) to keep their internal components (quarks and gluons) inside, and the only answer is 'distance', just like the distance between our galaxies, otherwise the quarks and gluons from one particle would intermix with the quarks and gluons of the adjacent particles, and the atom would 'dissolve' into a chaotic quark-filled gluon mess. [Just a bynote - the gluons could be seen as the equivalent of our 'dark matter' which keeps galaxies in our universe together.]

Now from our perspective, we see all of these particles as moving blindingly fast (indeed, at the speed of light) and being extremely small, where we cannot envision all the empty space between each tiny entity, from gluons to quarks to adjacent hadron particles, but to the particles/quarks/gluons themselves, they may see themselves as moving in slow, stately fashions relative to one another, much like our galaxies relative to the distances between them, where they have little effect on the internal unity of one other, much like our galaxies have little effect on the other's internal unities and dynamics.

The problem with the micro-galaxy concept is "uniformity". Groups of our galaxies are random in size and order, as is each individual galaxy is random in size

- no two galaxies are exactly alike (or at least there is an infinitely small probability).

Our atomic particles and subatomic particles seem to be of uniform types and sizes and properties, indicating that they are not composed of scores of unique star-like galaxies, which would give us an infinite number of particle sizes and shapes and properties, but are indeed irreducible units of matter formed by, and following, certain laws of our particular universe (which were determined by the nature of our particular Big Bang).

So micro-universes containing their own fleetingly (to us) intelligent micro-beings seems like a fanciful notion, but only if you discount the infinity of space, both outward (larger) and inward (smaller), where, at some size level, the uniformity of our particles break down and the randomness's and uniqueness's of another universe begins.

The intervening size distance between us and them could be called the 'quantum size buffer' – the relative size layers of matter that remained after our two universes completed all the possible interactions that they could (leaving the quantum size gap(s) between them), and where we have so for detected only the first few quantum size levels (atoms, particles, and subatomic particles).

What all this means is that such "micro-universes", and the fleetingly (to us) intelligent micro-beings within them, are still far, far (many times far) beyond our detection capabilities, let alone our interactive capabilities, such as intelligent communication, and like I said, such a gap can be represented by a number containing a certain number of zero's - say our instruments must have a power of 1,000,000,000,000 and we only have 1,000, then we are still nine zero's away from success (and we can celebrate every zero that we shave-off - moving closer to being able to actually test the theory).

The Stationary Model and Probability Clouds

Now a 'Stationary Atomic Model' offers an answer to such uniformity - where all those small units of smaller units are not composed of moving entities, but are stationary clumps. The problem is such a stationary model does not explain the probability cloud patterns that have been seen on particle/subatomic particle levels - the world of quantum mechanics.

Consider the following diagram:

Which Makes More Sense:

A. The electron, when absorbing and giving off quantums of energy, jumps from one distance to another, all while constantly in motion.

Electrons are 'allowed' only certain orbital distances. Do we see this anywhere else in nature? No. Is there any mathematical explanation for it? No - the math only deals with what it sees - the result, not the cause.

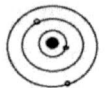

"I'm sorry, electrons, but you can only orbit at these certain distances - you are not allowed to be anywhere in-between, even when jumping from one orbit to another - you must do it magically - disappearing from existence at one level and reappearing from nothingness at the next."

Electron Cloud Pattern

Electron: "But what about my cloud pattern - how will I maintain that? You can easily see from the probability pattern that I already exist at all distances from the nucleus, and not just at a few prescribed orbits. You cannot explain it away by simply changing the term from 'orbit' to 'orbital'...!

Consider this preposterous statement: "In a hydrogen atom, the electron (in this cloud) is in its lowest energy state, or 'ground' state," and here is the referenced picture:

A probability cloud pattern!

I'm sorry - but an atom in motion is NOT in its least energetic 'ground' state! A 'ground state' IS depicted by my stationary model, however.

B. The electron, when absorbing and giving off quantums of energy, jumps from one STATIONARY point to another, the wavelength being the result of the distance between the two stationary points.

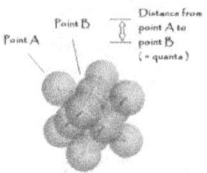

How it works: Electrons this close to the nucleus tend to stay there - since the attractive force is greater there. If you try to kick them up like dust they only make it to the next level up.

If you do manage to kick one up like dust so that it escapes the hold of the nucleus, another orbiting electron will have an empty space to fall into.

One question is, "How then do atoms bind through shared electrons in this stationary model? Well, the same question can be addressed to the orbital model - how does one tiny electron pull together two much more massive atoms? It would be like one planet holding two suns together. One would conclude that 'electron bonding' is a silly concept, and yet the model looks fine on paper, if you ignore the relative masses involved.

What an electron must be to bind two atoms together.

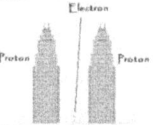

The 'recognized' relative masses.

The skyscrapers are two atoms, the man in the middle is an electron. Imagine the man trying to hold two diverging skyscrapers together - it is not going to happen.

In either model, stationary or orbital, it is held that one tiny electron has an electric charge equal and opposite to a much larger proton, and that the electron uses an electrical force to hold two atoms together. This is plausible, since we can see this at our level of existence - a small magnet has far more magnetic force than two inflated balloons, for example. Maybe this is a bad analogy, for the two balloons can cling to static electricity - an entirely different force; or maybe it is a good one - indicating that there is another force at work: but no, the magnet has far more mass than both balloons, so it isn't a good analogy.

The 'Cluster of Spheres'

Let's get back to this 'cluster of spheres' depiction of the atom's nucleus:

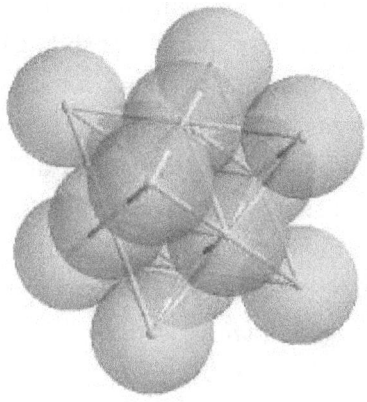

Since each sphere, a proton or neutron, is composed of spinning sub-particle particles (quarks) which themselves are composed of smaller active entities, this 'solid' sphere model is not correct, and therefore my 'stationary' model is wrong - for there are no 'solid surfaces' for the electron to rest upon - but then given the slow and stately manner with which particles interact (on their level), and the great distances between them, a solid surface is not needed - though they still could have them if their component parts are small enough (consider the earth being made of atoms and thus having a solid surface).

One question remaining pertains to the electrical charges.

To digress for a moment, looking at the spherical model, one thinks - what is to prevent the entire cluster (the atom's nucleus - quarks on up) from being scattered by the slightest perturbation, like so much dust in the wind? And, if they are attracted, what prevents all nearby atoms from joining in the fun and clumping together into one dense structure lacking the large empty spaces that seems to exist between atoms, no matter how dense the resulting substance is? In the Standard Model it is attributed to strong and weak nuclear forces, together with the electromagnetic force, with gravity thrown in. One does wonder then where static electricity (static cling) fits in, for it is a force to be reckoned with when objects become smaller, at least when trying to get plastic wrap to cling to anything other than itself, or trying to get a small piece off of your finger...!

Particles "At Rest"

You can see that, by their lack of mass and stationary fragility, getting an atom's particles to all 'sit still' is virtually impossible - one would need 100% entropy, or a temperature of absolute zero - and even then, the notion of 'stationary' would ultimately require a singularity.

As for scattering the particles, you certainly need to do more than just breathe on them - you can breathe on a steel beam and it will not fly apart. So there is a balance between 100% clumping together into a dense stationary super mass of particles (a singularity) and complete and permanent scattering of all particles at the slightest perturbation. Given Newton's first law of motion, these scattered particles would eventually collide with other atoms and scatter the other atom's particles in a chain-reaction; but, looking around at all the solid matter that surrounds us, we can see that that does not happen.

So one force is working to bring particles together while another is working to keep them apart, and the forces are so perfectly balanced that 'we' (multi-particle entities) can exist without our component parts scattering in the wind or melding us together into a singularity with our surroundings; and all this has been sufficiently explained and mathematically modeled; and it isn't the issue between a motion-filled orbital-based, particle-spinning atomic model and a stationary one, where the only spin and motion is that initiated by the external perturbations of the observer, as the act of observing kicks up the particle dust.

A Note on the Effects on Cosmology

So the residual energy of our particular Big Bang still exists, and is perpetually keeping the atoms in a disturbed state (and makes life possible - so far until the energy in our universe dissipates settles back to an entropic state of zero). Let's consider what happens then. The electrons find their 'rest' position, the atom's trajectory and speed slow enough that even the weakest attractions are enough to combine atoms into clumps, and the universe begins clumping back together again, ultimately into another singularity, until the pressure and heat build within it to the point where it explodes (or expands) again. How much matter is present is determined by how much was lost in the intervening time between singularities, and how much matter was collected from surrounding space from adjacent Big Bangs. So the system would not perpetually lose energy and eventually fizzle-out, where our particular singularity's universe would cease to function (like a machine which does not have perpetual motion - due to the second law of thermodynamics), since it could in theory pick-up new matter which expanded from other Big Bangs. This means that all those galaxies that are receding from us could be 'collected' by another singularity system while that system was still in its expanded state.

The Strong Nuclear Force and Fuzzy Neutrons

Particle physicists have theorized that there must be a special force keeping all the mutually repelling protons together in a nucleus along with all the neutrons, but they haven't really determined the source of this force. Well, I have an answer (a loose possibility) (at least a colorful conjecture) - 'Fuzzy Neutrons'.

You see, at present, the neutrons are just sitting there, not doing anything. Why did they decide to congregate there, if they have no attractive or repelling charge to move them around or bring them together? It make no sense - unless you consider my answer.

Think of the fuzzy neutron as one half of a Velcro pair, and the proton as the other half. NOW the strong nuclear force has a source - a physical interconnection, and the neutrons now have a 'purpose' - or their sticky fuzziness has led to the atomic nucleus and all the varieties of elements that we enjoy in the universe. A proton without a fuzzy neutron stuck to it is a lonely proton indeed, and hydrogen must be the loneliest element in the universe (multitudes of the lonely - poetically reflecting the human condition).

Second from last, the fuzzy neutrons fit with my 'stationary' model of the atom, and it explains why so much force is required to dislodge a particle from an atomic nucleus.

Last, this would explain why there are usually more neutrons than protons in heavy elements - there is enough surface area on a proton to accommodate more than one neutron, for example one neutron affixed to opposite sides (though what also might explain this are irregularities in proton shapes, say one is like an octahedral and another like a kidney bean, offering varying degrees of available surfaces for fuzzy neutrons to stick to.

The case against the 'fuzzy' neutron is that, according to current theory, if a proton loses a neutrino, it turns into a neutron, and vice-versa. Such an observation is hard-pressed to fit into the fuzzy model - what, in the fuzzy model, would make a particle obtain/lose a charge just by losing or gaining a part of itself?

The 'QUESTION'

Now, with all my conjectures (and anyone's) in particle physics (and cosmology and microbiology), the question is not whether any of these are possible - for, given eternity and infinity, any probable odds are rendered a certainty, but WHICH ONE is actual for us, when we ask, "Where did we come from?" and "Where we are in the grand scheme of things?" and "When are we in the grand scheme of things?" (along with the standard philosophical questions of "Why?" and the ever-present "Why even bother?"

Journal 60: Early Hominids – Leadership, Organization, and the Evolution of Challenges

Early Hominids – Leadership, Organization, and the Evolution of Challenges

The Backdrop

Consider the Oldowan tool-building site – a site at which early hominids built stone tools, a virtual factory that existed for nearly a million years...

So one hominid, standing on a slight rise, could see out across the plain at all the hominids squatting around the lake, building stone tools, and he thought, "There is such disorder here, and no leader – I can see right away that the survivability and strength of this haphazard gathering would be enhanced with a leader and centralized organization."

Og (to give the hominid a name, if prematurely – for named hominids did not occur for another few million years) knew he was right, but he also knew the hazard – that twisted individuals could bully and push and fight their way into the centralized positions and take control of the entire group (think 'Stalin' for a recent historical example, who, seeing the apparatus already built, knew that he could just murder his way to the top), though they were ill-equipped to lead in a wise manner (thereby decreasing the chances of the group's survivability) (and this in fact occurred, and the group died out).

Nevertheless Og took the chance, hoping to warn the group of the dangers before such a thing actually happened.

Our Og began by simply erecting a platform upon which he sat. "Build and they will come" was his reasoning.

So he sat upon his platform, and just as he suspected, the hominids, without a word, assumed he was set up as their leader by others who were wiser than they, perhaps by some god (and Og knew this was a card he could play), and they began to expect wise orders that would ensure their security and survivability, and they saw that they now had a wise one from whom to find answers, and they were happy, as the Bible would say.

Og found that, in this position, he began to have 'grand ideas' for the group, such as exploration and trade and mapping and mining and constructions of all kinds, and interacting, as a unit, with other groups of hominids in the surrounding areas and beyond.

He also had grand ideas for himself, and this gave him pause for concern – for it was possible that such egotistical aims could be detrimental for the group as a whole, and thus ultimately for their, and his, chances for survival; yet he also knew that the group needed a leader that they could be proud of – to 'show off', so to speak, to visiting competitors, which would result in their willing acquiescence, and result in a better guarantee of the survivability of the group, which includes all individuals in it, and hopefully for all hominids around, and for life in general (though such a true species Brain Age would not occur until our present times).

Concerning 'Empire'

The underlying philosophy here is that, if all hominids were of like mind, then harmony would exist, and the chances of survival would be increased. In this case, 'like mindedness' meant that all agreed on who will lead them. You can see how the twisted and perverted, seeing such a seat of power, would want to usurp it, even though they were not worthy of it, and even though it would be detrimental to the group's survivability as a whole, the twisted and perverted being shortsighted by definition.

Concerning 'Armies'

Our Og, and the group as a whole, felt the giddiness of the 'large cohesive, coordinated, armed unit'. They first experienced the advanced success of such an organized unit in hunting, and they knew nothing from the animal kingdom could challenge it. Then they were eager to test it against other groups of hominids, which in effect turned the cohesive unit into an 'army'. Now, just seeing two like minds looking at you is intimidating and awe-inspiring, but seeing dozens, hundreds, even thousands of like minds, well, this awesomeness just needed to be witnessed, so they set out on campaigns to impress others; and it was found that impressing others had it rewards – they discovered that just impressing others wasn't enough – they had to exercise this power and domination over them, which created tribute, which created unheard-of wealth. So you can see how 'conquering the lands of others' became such a popular sport – such huge armed arrays were indeed impressive, and they needed tasks and purpose to continue to exist (otherwise they would fall apart, placing the group in peril), which resulted in empire-building, which resulted in the collection of tribute, which resulted in wealth, which resulted in survivability, at least initially, before corruption set in.

Concerning the Challenges

Now, just as the hominids progressed socially and technologically, so too did the challenges they faced – as if the challenges also evolved; and, in essence, the challenges did – the hominids found new challenges every step of the way on the road of progress, and this became a general rule of evolution – as one evolves, so too do the challenges one faces.

The Females

How did the females handle all of this? They thought the increased security was a great thing, as was the new wealth; but they wished it could be attained a little less violently – their modus operandi being cooperation and understanding, or at least domination through suggestion – but males had little patience (or capacity) for this, not when the excitement of running around beating things (and resisting hominids) with their shiny new stone tools was an option! (and you can see that, two million years later, male hominids have progressed little).

As a result the females concerned themselves more with mini-goals such as rank and status and refinement and lineage and fortune and position and family, that is, once basic needs were met and idle time became available.

Journal 48: The Artist's Roll in the Advancement of Science and Technology

First a short note: I've always held that anything 'technical' should be presented in any form other than sentences - for what are sentences but vehicles for transmitting mental images and relationships - and they are very inefficient vehicles at that!

So, for the clearest presentation (to ease the forming of the intended mental images), here is an illustration:

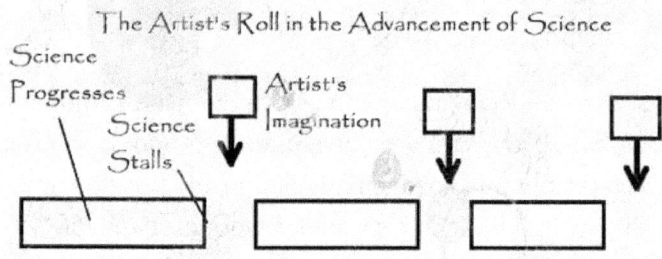

The Artist's Roll in the Advancement of Science

Science Progresses

Science Stalls

Artist's Imagination

What you see is science, for example, 'progressing', and then reaching a 'roadblock', a stumbling point, a 'wall' - in effect, it stalls, and this is where imagination is needed; and who better than a natural artist to supply an ample amount of imagination? So the artist contributes wild imaginings, where one just may be the spark needed that renews progress in science by providing a new path for inquiry, or, as in the case of technology, a new path for development.

Example

Here is an example of an artist's wild imagination offering unheard-of possibilities...

Juno G and Her DNA-Based Cube Cyborgs

While onboard her starship, Juno G answered my question about how she created the cube cyborgs (she gave a dynamic demonstration through music) (music to follow).

She began with an explanation of the DNA structure she designed at their core, which is a cube-based DNA system, where each cube is referred to as a 'frame'.

The Cube
or "Frame"

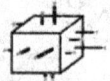

Inner Attractive Force

Interlocking Connections

The DNA
Strand

"Magnetized" tips - red and
green representing each
opposite pole,
where opposite poles attract.

Attractive Forces

Outer   Outer

Inner

How the
DNA
Strand
Duplicates

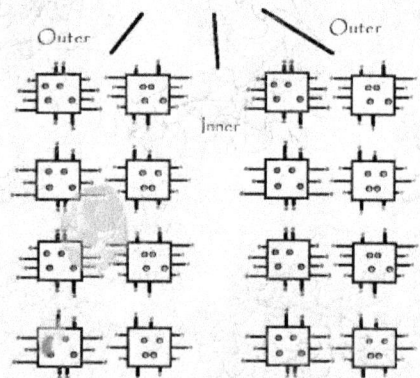

Each frame attracts an outer frame. It takes two 'outer' attractive
forces directly opposite to one another to 'pull apart' an inner
attractive force.

In a chaos system, mutations (new 'blooms') occur when duplication
occurs before an entire strand segment is fully formed, and this gives
rise to 'evolution'.

How It All Works - Chaos Systems

Levels of Attraction

In a chaos system, 'attractions' are used to 'make things happen'.

The Weakest Level of Attraction – is bond-to-bond, which requires close proximity.

The Next Strongest Level – is frame-to-frame, which works at a larger distance, and which creates a conglomeration, or higher concentrations, of likely interacting frames, which increases the odds of the intended frame-to-frame encounters, meaning better chances of actual frame-to-frame couplings, and decreasing the chances of a system with no dominate encounter type, where a system with such random encounters, with none dominant, as in a "shake-and-bake" system, would only result in a tar-like sludge comprised of all possible combinations and the results of their reactions.

As frame groups increase in size, the associated attractive forces increase, first between groups of frames, and then groups of groups of frames, until the final structure is realized – the final structure being ultimately shaped by the final overall attractive force of the structure. You can see that the weakest (smallest) attractions are quantumized, while the larger attractions become more and more randomized.

DNA Blooming

This is how DNA strands become distinctive parts of a system, for example, a liver (which is a component part) of a system (you). An analogy is a string of various connected flower seeds (the DNA strand). Once nourished and allowed to 'bloom', they become an interconnection of various mature flowers (the final 'system'), where they then all 'work together' in 'existing', such as an animal, such as you, or in this case, Juno G's cube cyborgs.

Each 'bloom' is initially composed of a certain pattern of strings, connected in only two dimensions; but when they begin to 'bloom', they connect on all sides, 'growing' into their final cubic structure within the system, where they then begin to perform their given function. It has been noted that when cells combine, they can begin to perform higher, unforeseen functions (such as becoming your liver and performing that function in the system). There is only one possible outcome 'from strand to structure', based on the 'code' of the strand, barring mutation, which can occur at any stage, and which thus have various altering effects on the shape and function of the final structure, begetting new shapes and functions, the best of which continue, comprising evolution.

The Dangers Involved – Cube Cyborg Dominance

So, in summary, I asked her how they built the cube cyborgs, and she showed me that they began with 'seeds' of code, and just let them grow and evolve from there.

I asked her of the dangers involved – say, a runaway evil cube cyborg race, and she said sure – they will go through various unpredictable developmental phases – from mindless animals acting on instinct to semi-brained evil entities to fully-formed enlightened, beneficial beings – and that yes, there were the initial potential dangers in each phase up to the final beneficial stage, and that, for her survival, she had to place faith in her own ability to react and adapt to changes, knowing that she had a few billion years head-start in evolution – which gave her the advantage. Another factor was their relationship - if she just used them as slaves so she could become stupid and lazy, then yes, they would come to dominate her; and if she outright abused them, well, she would deserve whatever vengeance they could muster.

All this is a moot point now, but it was important before an adequate universal philosophy evolved, which created understanding, tolerance, and harmony in diverse beings.

She told me that one Numi Earl Grey developed it on Earth, and, although people on earth rejected and ridiculed it, she adopted it and found it adequate for guiding herself and all that she designed which had self-evolutionary and learning abilities (meaning being able to understanding, and thus be guided by, such a philosophy).

Journal 43: DNA Frames in Chaos-Based Systems - A Nice Way to Connect and Part Ways

or, "Introducing Chaos-Based Engineering"

or "How Life Operates on the Molecular Level"

or "Just Another Line of Original Thinking by Numi Earl Grey"

I was trying to envision how DNA split, and how proteins were made, and I concluded that it must be a chaos system that operates on random chance encounter. So I then tried to imagine the detailed mechanisms at work in a chance encounter. I came up with a simple solution – apparently not as complex as reality, according to recent discoveries, but it is a nice way of doing things nevertheless.

You have the AT and CG nucleic acid pairs contained in pronged frames, let's say cubic frames for illustration, with let's say four prongs per side, with little magnets at the ends (really electron action at the molecular level), each prong a different length. Each of these cubic frames carries (and protects) a single nucleic acid, A, T, C, or G. For the proper frames to connect together, they must have four prongs with complimentary lengths, or the attractive force needed (all four prongs touching) will not be great enough to keep the two frames together – say if less than four prongs matched – the attraction would not be strong enough, and the frames would drift apart.

So first you have the DNA strand, all kept together, from top to bottom (which creates a dilemma – you would need a lot of prongs to contain the code for the right sequence (math below)), all keeping the AT CG nucleic pairs together side to side, on the 'inside' of the two cubic frames, say an AT pair. To separate, you would need each frame to connect with another frame on the 'outside', one T and one A (having the matching nucleic acid). Having two outside frames would generate enough force to cancel the inside 'holding force' of the inside two frames, which would then 'separate' and drift apart. You would think that the problem here is that, since all forces cancel, none of the frames would stick together, and everything would drift apart, but no – only the inside connection has two opposing forces acting on it.

OK! (I got through that crisis!) Now you have two strands from one, but 'inside out', with the nucleic acids pointing outward and not really 'connecting', making for a non-functional strand. Each strand would then have to attract 'outside mates' which would then split the strand, and now you have two new regular

DNA strands exactly like the original.

The problem with this model is the very chaotic randomness that the model was supposed to function in. What is to prevent sections of strands (or individual frames) from being 'pulled apart' (really 'drifting' apart since the attractive forces of the two inner frames would be canceled by the opposite forces of the two outer frames) before the strand is 'complete' lengthwise? The only answer, and not a plausible one, is that the strand cannot separate until it is complete lengthwise, but, from a mechanical point of view, this doesn't hold – there is no reason for things to be that way. So these frames would never have the 'chance' to form 'complete' strands lengthwise, or even protein code sections lengthwise, each with many millions of pairs, before something drifts away. Too bad – it is a nice simple system otherwise!

As for creating proteins (portions of the main DNA strand), the separation of those would be done independently, section by section, but even this flies in the face of a chaotic, random system, where incomplete protein strands would be connecting and drifting apart at random, unless you could survive on the extremely small odds that a complete protein strand is created before a piece of it drifts apart, in which case you would see a lot of 'debris' in the cell. Same goes for a complete DNA strand, though with an astronomically greater amount of 'failed debris' floating around, and astronomically smaller chances of forming a complete strand, which relates my original premise – that we do in fact exist on such astronomically small odds, owing to the astronomically large number of molecules involved (and which leaves a lot of room for 'improvement', just to mention that).

I'll include diagrams to aid in the illustration of all this to help clarify things, and maybe doing it will shed new light as to whether this system would work or not (even though it may not be the way things actually work in our biological systems, it may be suited for artificial, chaos-based systems.

Strand code math: Let's say your strand contains a billion pairs lengthwise. How many prongs would you need to create a billion different combinations?

Answer: You could do it in binary, where you would need 30 prongs (2 to the 30th power = just over a billion), and you would only need prongs of two different lengths (representing the 1's and 0's of a binary number system).

Properties of (Creating) a Chaos-Based System

First you need chaos. Chaos involves 'things' in random motion. In living cells, there are many trillions of these 'things' (atoms), and it appears that the cell manages to keep everything in constant motion internally. This can be done in several ways, with heat in a gas or liquid, or in liquid that is kept in motion, which results in chance encounters. I say 'chance encounters' even though molecule 'A' may be attracted to molecule 'B' - their attraction is limited by distance, speed, and trajectory, so you cannot 'order things' as in ordered engineered designs.

So, you 'throw' many 'things' into this 'motion base', and you have your chaotic environment in which to let your system 'run'.

Many interesting statistical relationships will exist. Some less obvious examples would be, the less variety you have in your 'things' ('elements'), the slower your system can operate (using less energy), because the odds of chance encounters will be higher, given a fixed number of elements. If you increase the variety of elements, the less likely the chance encounters you want will occur, and you would need a faster system to increase the odds of the desired chance encounters ('success') (all systems being measured in a standard time span). The more energy applied to motion, the higher the success rate. The smaller a system for a given number of elements is, the higher the density, and the higher the odds.

Existing Example

Consider this statement about the internal components (which include the organelles, nucleus, and cytosol), of living ectokaryotic cells: "Often, cellular compartments are defined by membrane enclosure. These membranes provide physical barriers to biomolecules, where transport through such barriers is controlled in order to maintain the concentration of biomolecules within and outside of the compartment."

In a 'chaos system', what these membraned compartments are in effect doing is increasing the odds of the desired chance encounters to occur, thereby increasing the odds of functional success.

This example in fact gets right back to the reason I considered 'chaos systems' in the first place - trying to figure out how life works on the molecular level.

Chaos Systems and Evolution

In response to a statement denying evolution (in favor of creationism), I responded: You may be right, but you are needlessly denying evolution - you do not need to deny evolution - for evolution is a great 'design tool'. Take this experiment (which has been done): Say you want to 'design' a certain class of proteins that behave a certain way. To mathematically create them by artificially constructing their DNA is currently beyond our capability, yet it has be done with evolution in only a few weeks, where, out of a chaotic mixture of amino acids (which form proteins), the 'first generation' of proteins produced a few 'winners'. After filtering out the debris, the winners are allowed to duplicate, and some 'children' were even better at the behavior, and in larger numbers. After around 20 rounds of 'evolving and filtering' you had a lot of proteins that were very good at your desired task - without any mathematical engineering whatsoever (which would have been astronomically difficult - which is why we cannot do it that way yet, if ever, and which would be impractical when we have evolution as a 'tool'), which is why 'evolution' is a great design tool - for living things, or any design based on a chaos system.

Your statement also illustrates how any science and engineering can be placed under the umbrella of any religion - for "because God made it so" can be used for the answer to any as yet unanswered questions as to 'why things are', but to do that would be to turn off our minds, which would keep us ignorant as to how our universe works, which would lessen our chances of survival. So your philosophy is a philosophy of 'ignorance and death'.

Conclusion

What I'm introducing is the concept of chaos-based engineering - which may be the best way to affect the best solutions; and it may very well be the way we are (biology is) 'designed', or have evolved.

Journal 38: My Chaos Theory for Life (on a Microbiological Level)

In our universe, nature operates on randomness and chance in a chaotic system, so it follows that all the biological functions within us run on randomness and chance, and are basically chaos-based systems - though with 'life', the odds were increased with haphazard developments such as cell membranes, which created self-enclosed environments where the odds of nutrients finding their mark are increased by increasing their quantitative density inside the self-contained environment (as compared to the outside).

For this theory, I am also going to assume that all religions are make-believe, and there is no guiding hand in the mechanisms of the universe or the life in it. What is left then? We are right back to chaos - all the blind forces acting on matter, and we can look at life happening and evolving in terms of probabilities.

Atomic physicists, cosmologists, microbiologists, and other intrepid explorers trekking into the frontiers of knowledge would love to find ordered beauty in nature, a 'beautiful universe' within and without, but nature is more like a chaotic conglomeration of chance, and we need understand the universe we've attained consciousness in - meaning the state that we exist in, what we have to work with, and what we can do with it.

Consider these current statements from microbiology:

"An AUG is also used as a start codon, which 'tells' the ribosome where to start translating the base sequence into a protein…" There is another statement which describes one molecular entity 'telling' another one 'where' to go; and consider this video:
http://www.youtube.com/watch?v=GigxU1UXZXo

You can see that the micro components of a cell are depicted as having
will – that is, efficiently going around with a specific task in mind,
and also knowing where to go to carry them out.

The truth is, it is all driven by chance encounters with extremely small odds of occurrence. What makes it statistically successful is the sheer quantity of molecules involved (in the trillions), and the speed at which things happen at the molecular level (for protons and electrons, at the speed of light, we are told in physics).

Take the AUG statement above. It would appear that, given one AUG and one ribosome, the AUG, being endowed with some mysterious sensing apparatus, will smell-out the ribosome, find it, and carry out its task, but that is not the case. It will take a million, maybe a billion, AUG's and ribosomes for just one chance encounter to occur, and, statistically, this is enough for the cell to continue to function. This means two things: that one, on the molecular level, the operations of the cell are based on chance encounters with very small odds of occurring, and two, there is a lot of room for improving efficiency, which is possible given our 'brain era' of existence. What it also means is that 'life' cannot exist without wildly vast numbers of active subcomponents (read the trillions of molecules in our case and their active electro-chemical interactions). This seems self-evident, but it is rooted in chance encounters and sheer statistics.

Why do things happen the way they do when a chance encounter occurs? It would appear they are due to molecular material laws of nature rather than molecules that 'think' and have will and a purpose (anthropomorphic molecules) – for example if a certain AUG encounters a ribosome, a certain protein will result – all blind, rather than intended, and all driven by natural forces, and all by very slim chance encounter made possible by sheer numbers.

Why are things the way they are? That is what exists at the moment in a blind, trial-and-error system of evolution. We, for example, are a trial, and whether we are ultimately an error or not is still to be determined (given the 'ultimate goal' of an 'ultimate being' when taking evolution to its ultimate end).

Other Considerations

Thinking further on this, you would then wonder if all 'chance encounters' are benign (and not malignant), and if there are only a certain variety of encounters that actually result in anything at all (a subsequent action). Again, why things are the way they are can be attributed to the chance encounters and trials-and-errors of our life's particular evolution (over billions of years, remember), all played out along the lines of material physics (physical forces) (which underlie chemistry) (which underlies biology).

Doing the Math

I'm not a physicist, and the math is in the area of physics, and it involves thermodynamics and energy, but I can say what path to take: you would begin with 'what is' - that being us, and the energy we expend while existing. You then take that quantity, and balance it against the number of molecules at play (within us), to derive a percentage of chance encounters necessary for us to continue to exist. For example, let's say we expend on the whole one joule of energy every microsecond. We then calculate how many molecular reactions are needed to create one joule of energy, and then out of how many molecules are present, to derive the percentage of chance encounters needed.

In short, beyond field strengths and shapes and attractions and repulsions, and beyond thermodynamics and kinetics, the math of life will be in probabilities - the probability of the desired molecular encounters and chemical reactions occurring.

What is also intriguing is taking the final product, such as us, and working backwards to determine the condition of chaos that we originated in.

Improving Our Biomolecular Efficiency

We began in nature, and continue to exist in nature, as a compendium of chance encounters, all after a long series of evolutionary trial and errors (over eons now). Now that we are in the 'Brain Age', we can begin to tinker with things, with an eye toward increasing biomolecular efficiency (by increasing the percentage of the necessary chance encounters, whereby we would internally be able to produce more energy), or, conversely, decreasing the number of chance encounters needed to 'live' as we already do (needing less energy to exist).

So the question is, not whether life uses chaos-based systems, but how, at each evolutionary stage, the odds were improved (and porous vesicles, that is, self-contained systems that increase concentrations of desired components are usually the answer - porous so the contained system can still interact with the surroundings to regulate the inner workings - to take in needed material and eject waste).

On Eating (and Digesting) Porousness, and Hunters and Prey

It is the taking in of needed materials into our self-contained system that is the secret to increasing the odds of the desired chance occurrences occurring. It is why we 'eat', and the reason we need to 'digest' - to break 'food' down into the required component parts.

It is why cells have porous membrane vesticles, and why multi-cellular organisms contain all their cells in a none-porous membrane vesticle (what we call 'skin'), and solved the problem of none-porous intake and expulsion by creating an intake point (the mouth, to gather the required material) and an expulsion point to eject waste, Mobile organisms increased their odds of obtaining the required material, and thus flourished (at least those with the mouth in the front and the expulsion point in the rear!). Since not all material is beneficial, indeed, most were 'non-nutritious' or 'deadly', organisms that developed the ability to select flourished - which created the 'hunter-prey' system we see today - because, it just so happened to turn out, the 'best material' was contained in other living life forms - plants for 'prey', and pray for the hunters (and now that we've entered the 'Brain Age', we can synthesize our own non-living 'food', and thereby free ourselves from the monstrosities of being hunters, even though life has evolved to minimize the pain of the prey through 'shock'.

On Our Senses and Their Origins

As to our five senses, the hungrier we become, the more heightened our senses become (as humans, we over-eat to deaden our senses to the misery of the current human mindset). With heightened senses, we increase the chances (if primitive) of our finding more nutrients to keep the probability of life-sustaining molecular interactions high within our cells.

The purpose of our senses (hence the reason for their evolutionary survival in a chaos system) is to obtain nutrients. Those organisms that developed vision, smell, touch, taste, and hearing gained advantages in obtaining nutrients, just as mobility, size, intelligence, and strength did.

The two-sex system (and the advantages thereof - meaning promoting cooperation and combining diverse sensibilities) could not have developed without the securing of nutrients (giving us idle time), and senses made that possible. The two-sex system reduced violence and promoted compassion and cooperation, and improved care for the next generation.

Improvements Through Chance Evolution

Chance also plays a part in evolution, though the testing of chance mutations that improve an organism's senses were, and are, hunger-driven. Take the eyes, for example. They were not tested until hunger set in and more material needed to be ingested to keep the odds of the necessary internal chance encounters high. That organism that had vision had an improved success rate, for it learned that if something 'moved', it probably contained a high concentration of the right materials. Those organisms with eyes close to the mouth had the highest success rate, pushing all other organisms with inferior variations to a quick extinction.

Answering The Shake and Bake 'Tar' Dilemma

It has been argued that a living system cannot operate on chance encounters, because all the possible variations of chance encounters combine to produce a useless tar-like substance. Well, our bodies DO produce such 'tar', the 'tar' that our bodies produce is brown, and we are constantly getting rid of it.

Yet, one may wonder how likely our chance encounter system is, and the answer is as likely as our ever existing. Evolutionarily speaking, advancement embodies increasing the odds of the desired chance occurrences, meaning increasing the efficiency of the system.

On Failing to Eat (or Digest)

if we do not eat, the needed material diminishes in numbers and thus in concentration, and the odds of the desired chance encounters occurring diminishes. If we continue not to eat, the odds become so low that we begin to 'die' - our 'body', to give our self-contained vesticle a name, may begin to look for the numbers from within, in which case it 'eats itself alive'. Continue not to eat, or lose the ability to 'digest' - or break-down 'food' into the necessary component parts, and the ability to keep the odds of the needed 'building blocks of life' sufficiently high diminishes, until the odds spiral downward, out of control, until we are actually 'dead' - meaning the governing functions cannot function any longer, and the necessary number of chance encounters to 'sustain life' have no chance of recovering.

On the Development of Coordinated Systems of Space and Time

The quantifying of space and time facilitated the coordination in securing nutrients, and in orderly reproduction ('society') and in combining individual efforts in solving larger problems (again 'society'), which becomes possible in the Brain Age, where complex, specialization-based societies are critical.

So our senses of the past, present, and future are all based on securing nutrients.

The Mystery of Death

Why a chaos system should gradually fail and die is still a mystery. There could be many reasons, and the most pertinent here is that perhaps the whole model of a chaos system is false.

Summary

Life is a system born and evolved in chaos, according to the physical laws of the universe (not all of which are perceived or understood yet). This is the 'why'. Now that we have entered the Brain Age, we have the opportunity to find out 'how'.

Approaching 'life' from this 'chaos and probability' theory may help us understand 'life' better. Taking this approach may explain such things as how extremophiles and viruses work, and it may be the path to take in fighting diseases that are still beyond our means to prevent and fight them.

Post Script

Well, happy tinkering, oh denizen of the Brain Age (of evolution - another example of an adequate philosophy updating itself in the face of new scientific discoveries, a chance encounter itself (chance encounters, like trial and error, contributing to the evolution of a system), the 'brain age' being just another trial-and-error, chance-by-encounter natural phenomenon, which may or may not work out, being trial and error and working on chance encounter - so try not to break anything - we need to keep the odds high.

One final note: Say some or all my notions ARE true and we come to understand how life works just a little more, then let's not just 'smile broadly' together, let's jump up and down in ecstasy together! (it will be a good excuse (or 'opportunity') to burn some calories (increase the odds) and gain some strength (so we can increase the odds of gathering the needed material and digesting it adequately).

And remember, exceptions are in keeping with probability and chaos.

Where does Math Fit In Beyond Probability?

In a chaos system, such as what life originated in, exists in, and operates on, math can only describe what is and what has been, it cannot predict the future.

Why is There So Much Useless Code in DNA Strands?

It could be that those areas are not useless and do nothing, we just haven't found their uses yet. To give an example, DNA code may not only need to know 'what' is needed (for example to create a liver in a human), but how many cells (or cell divisions, if this is the right way to look at the problem, and we do not even know how to look at the problem yet, which is sometimes half the battle) it will take.

Summary of How Atoms and Molecules 'Work' in a Chaos System

So how would atoms, and then monomeric and polymeric molecules like amino acids and then proteins, 'work' in a chaos system? Whether two atoms combine or not is be determined by their three-dimensional shapes, their sizes, and their electrical positive/negative patterns. This would account for why some atoms combine and others do not. What causes them to combine is determined by proximity, and also sometimes trajectory, and also sometimes speed when more energy disturbing the atoms is needed (to make the atoms fly around farther, and in more random directions (bouncing off non-compatible atoms), and faster, for example when a catalyst like heat is used). The same holds true for molecules, macro-molecules, and all the way up the scale – all the way up to human love, perhaps, and why we do not stick to rocks.

A Note on the Cosmological Relationship

So energy is really the driving force behind life. Once the universe ceases to have energy (gradually settles back into a complete state of entropy), matter will drift back together into a singularity, and if enough matter is collected, internal forces (pressure, heat, electrical) will cause it to expand again into its expansive state - what we call a universe.

On Viruses and Vaccinations

This chaos system would be why using dead viruses works as a vaccine - 'shape' is the only criteria, which triggers the production of the necessary complementarily-shaped white blood cells to attach to and neutralize the virus.

Hitting the Hypothesis with More Challenging Questions:

How Does Epigenetics Work in a Chaos System?

The genetic tags which turn gene sequences on and off merely change the shape of the gene, creating a condition where the histone proteins that normally interact with the gene sequence can no longer interact (interactions being based on compatible shapes).

How Does a Living Organism 'React' to Its Surroundings?

If 'reaction' involves 'restriction', say, of a gland via muscle contraction, then the smaller size of the gland will mean a higher probability of producing the needed stimulus. As it is, glands store their over-productions, which are then immediately available, being 'squeezed out' by reactionary constrictions of surrounding muscles. The glands then gradually refill the readily-available stock.

The Next Step in Evolution

Since we are intelligent enough to manipulate evolution now, the next step is to move from chance/chaos-based biological systems to 'enhanced' (where the chances are increased even further than what membranes gave us); and when we reach a certain level of 'enhanced', we can call it 'controlled'.

Where Does Death Fit In?

You would think that such a chaos-based system, once begun, would continue on forever if adequately fed, but a living entity does not continue on forever. Why? What gradually fails in a chaos-based system? Do our membranes become leaky, or alter their composition where the needed reactants no longer penetrate? This is still an unexplored area of investigation, and the investigation is still in the speculative state (where creative thought and inductive reasoning come into play, to offer new possibilities worth investigation).

What about Entropy?

The important scientific question here is, "Is life anti-entropic, arising from nothingness, or is it just riding the universe's energy wave down to complete dissipation, where it will end?" and the important philosophical question here is, "If life is just riding the wave down, can intelligence overcome the ultimate end?" (though it isn't the ultimate philosophical question, which is "Why bother?").

The 'Odds' Argument Against 'Chaos-Driven' Life and Evolution

I just came across an argument against chaos as an environment for life - the argument claims that, since the odds of obtaining a complex chain of proteins is 'hyper-astronomical' (the number being larger than all the hydrogen atoms estimated to be in the universe), then the chances of the protein existing is effectively 'nil' - meaning there must be a 'driver' for life - a natural force that increases the odds of such patterns to come into being, like a carrot on a stick leading life as we know it on in a certain developmental direction. This is nothing more than another "light through an interstellar ether" notion.

This is a fallacious argument in that, no matter what the odds are of creating a specific protein chain, here they are - they exist. The argument inherently states that since the odds are so hyper-astronomical, then life based on such proteins will not occur. This assumes that the proteins were a 'goal'. If they were a goal, then yes, the odds would be nil in achieving that goal; but our proteins were not a 'goal' - they are what happened, and 'odds' do not apply here - life in a chaos is not a 'predictive' issue, it is a hindsight issue - it is a 'results' issue - there is no prediction or planning in a chaos environment - for it is THEN that you will encounter those hyper-astronomical odds.

In summary, you do not look at what already exists from an 'odds' perspective, and say that it cannot exist because the odds were too high - whatever the odds are, life - even based on such unlikely proteins, exists - here is it - the hyper-astronomical odds-beaters.

Just to note, you could also say that such a universe as ours that gives rise to life has hyper-astronomical odds against it - which means it should not exist - yet here it is, and here we are contemplating it (some more successfully than others).

So the 'odds against' argument is false in that it is an 'all or nothing' argument - either the hyper-astronomical odds will ensure life (and the underlying protein chains) will never exist, or exactly such life will come into existence in spite of any odds (which is in fact what happened -and we are witness to it).

So, in the end, you could turn the 'odds' argument completely around and state that the odds were 100% that such life WOULD exist.

Molecular Plasticity and Complexity

A note: One thing that would 'increase the odds' are 'flexible molecules' - molecules with a plasticity of shape, where they can alter themselves (to a degree) in order to couple with another molecule in a life-friendly (life-supporting) manner.

Another factor is molecular complexity - where even partial interactions 'count' toward the perpetuation of life.

A Note on DNA Replication Accuracy

To remain reproductively accurate, there may be many sub-steps on much smaller scales - redundant enough to allow for many mistakes, and still have an accurate end result on a higher level.

A Note on Protein Fold Vibrations

Protein folds, by vibrating randomly, create (blindly) increased odds for interaction (finding a matching compliment). The question is, what are the maximum odds sufficient for life...? One in a billion are not too high.

The Concept of 'Force Patterns'

Considering the astronomical improbability of complex life existing, some say that it is not possible for us to have originated from a state of pure chaos.

They are on to something, though totally wrong. It may be that 'pure chaos' cannot exist for very long. What begins to happen are 'patterns', which have 'forces' (such as molecular attraction) associated with them. A 'successful' pattern is one that arrives, then survives, then grows and multiplies - simultaneously increasing the probability of that pattern occurring. If one such pattern was the basis for life, bingo - here we are, growing more and more complex as successful patterns grow, all driven by the associated patterns of force, which attract/drive accumulation and thus growth.

Such 'force patterns' do not have to be limited to singular and static - say from one certain molecule. They can take on the nature of 'sequences' - certain sequence patterns, such as biological metabolism, where the 'successful' ones will stick around and procreate.

Good bye, pure chaos, hello chaos inhabited by Pattern Forces. Humans can be considered Pattern Forces with eyeballs.

Subsequent Passing Thoughts

Perhaps the probability of consciousness forming from physical chaos is affected by the density/force relationship of the building blocks involved (in our case 'atoms'). Given a great enough density and a large enough force per unit then interactions will happen and a physical structure that issues consciousness results (perhaps inevitably)

The Impulse to Assemble

It has been said that nature has an impulse to assemble, such as lipid compartments and snowflakes. We can extend this - that life is comprised of those assemblages that assemble in pairs - which gives them the ability to 'reproduce'. Extending this further, it is perhaps why we have a lot of symmetrical features (left-right equivalencies such as eyes, ears, and limbs) - though in our case we cannot split right down the middle - we 'solved' this (or nature did via chaos and chance) by reproducing on a molecular level (sex cells and chromosome pairs).

So 'life' is the progeny of those molecular assemblages that assembled in pairs, then took advantage of lipid membranes to create a high-probability-creating enclosure for life functions to occur (in a random, chaotic system)..

End Note

Whether right or wrong (in the end), in the meantime this 'chaos theory' (as with any of my theories) may serve as another 'perspective tool' with which to consider the yet unknown.

Journal 31: It Follows... (Science Paradigms)

Time Travel

According to Einstein's theory of relativity, time slows as speed increases. If this is true, then 'time travel' is possible – but only in the forward direction –as you 'travel around in circles at nearly the speed of light', you slow the time around you, while the world continues on at its quicker pace. So, for example, you can 'travel' in circles near the speed of light for a period of a hundred years, where time nearly stands still for you, then slow down and return to earth, a hundred years in the future.

But there is a paradox – since your time-slowing is only relative to theirs, their time will have slowed relative to you (for to you, they will have been moving at nearly the speed of light – because, according to the theory of relativity, you still seem normal to yourself as you travel at near the speed of light – time doesn't 'actually' slow down), it is mere equation balancing (so we need a new equation); so no one will have gained any time.

As for longevity, it follows that traveling near the speed of light will increase your lifespan, and it will make traveling to the stars possible – though it may take, say, 4.24 light-years to get to the nearest star - Proxima Centauri (the little red dwarf of the trinary system), it will be almost instantaneous since time has slowed to a near standstill for you (and your physical being). How about that! Unfortunately, the first "Model-T" speed-of-light craft may take your entire lifetime just to get up to speed (and then another just to slow down). Traveling would be quite safe – you accumulate mass as you pick up speed, so at those speeds, you would be so massive (meaning 'dense') that you would obliterate (or poke a hole through) anything that got in your way (which may account for solar flares – light-speed travelers passing through); and since forward-scanning radar would not work (you would be going nearly as fast as the radar waves. which would not be able to return any information beyond a few inches), being massive (massively dense, that is) would be critical...

Turning Energy Into Mass

According to Einstein's theory of relativity, objects accumulate mass as they speed up. So, it follows that if you want energy to turn into mass, you merely have to 'speed up' the energy. Cosmic-speed convection currents (approaching the speed of light) would do this, for example. Where did the heat energy come from initially (assuming energy existed before mass) to cause such 'convection currents' (that occur between two points of differing temperatures)? Perhaps from space stresses – 'ripples' so to speak, or electrical charges, causing a sort of 'lightning bolt' to rip across an entire universe, and somehow creating a particle or two from the energy within it.

So where did all the material forces (gravity, magnetism, electrical, heat, and nuclear, for example) come from? Again, 'space stresses' – space rippling and bending as it interacts with infinity, perhaps as it 'grows'; or from electrical-like charges, probably like static electricity, which is very clingy...

So what existed before space 'grew' into infinity? Time.

So how much 'space' and time did it take to create one atom? A lot! But it depends on how much energy was contained in that space. I want to say it took a space the size of our present universe to create one atom, but that may be a bit much. We can calculate how much energy an atom is made of, but we don't know how spread-out all that energy was 'initially', or how long it took to coalesce into an atom; and, as with that cosmic lightning bolt, it could have been instantly.

So how much time did it take to create the first atom, and then the next, and then finally all the matter we can see? A very long time! (relative to our lifespans).

ON THE SPEED OF LIGHT

If $e=mc^2$, then $c^2 = e/m$, so it follows that if you change the energy to mass ratio e/m, then you change the speed of light. In nature, e/m might be constant, but that doesn't mean we cannot manipulate it (in order to 'reach the stars', for example)... (or, 'just because'!).

Paradigmiclical Illusions and Shifts

A paradigm is a model of reality, and the hazard is that they can sometimes be mistaken for the actual definition of reality, and when the real one comes along, it is rejected, people clinging to the fake model! (Just ask Galileo). It can also be a paradigm-iclical (or add your own creative suffix) illusion, in which the same reality can be modeled (seen) in different ways, (and in the rabbit-duck illusion), all of which are wrong, or, at best, partially right, where they can then be partially used (within a certain probability) to make accurate predictions - that is the end result of science anyway.

What I'm saying is that above, I am giving you a paradigm of how energy and matter came into being – it is non-mathematically based, which makes it an 'effective' (speculative) theory, rather than a physical theory (based on measured observations) or a mathematical theory (based on mathematical models). Speculative is the easiest, academically speaking, but it has the broadest potential and can make connections and see cause-effect relationships that the others cannot; and it is the most fun.

Any theory beyond present mathematically-based science becomes polysemic, and which 'seem' (from poly-se-mic) is right (or partially right) is anyone's guess. The conflicts that arise in this area between people are, just to note, not a cause for anger.

A major paradigm shift (mindset shift) (mind-frame shift) (model of the world shift) (a person or society's worldview shift), meaning on a society-level, is a media-worthy event. Otherwise it shifts slowly and imperceptibly.

On the Joy of Thinking

Those who love to think often come up with humorous observations; hence, humor equals a love of thinking. You can say 'intelligence', but intelligence involves more than just thinking! So what I'm saying is that I may not be right, but I am thinking, and there is a certain amount of humor.

22: Solving Einstein - or The Joy of Thinking

or, "On the Scenario of Sub-Particle Galaxies"

or, "A Demonstration in the Joy of Thinking"

or, "How a Poet Solves Einstein"

or, "The Fun of Imaginings"

or, "Bringing a Smile to a Thinking Person"

or, "A New Physics of Life"

or, "Nested Universes"

Solving Einstein's Unification Theory

I was initially going to say the answer was 'static electricity', considering how material 'clings' ever-stronger as its size decreases, but this doesn't account for the other independent forces, such as gravity and magnetism, nor does it answer the question of why mass attracts mass.

Today we 'know' the thermo- and electro-dynamic (chemical) principles of atoms and molecules, which addresses mass and energy, but only on our quantum size plane.

It is said that Einstein spent the latter half of his life trying to 'unify' his theory of the universe with that of quantum mechanics, where energy was found to be emitted in 'quantum packages' rather than as a steady stream, and the electrons moved from valence ring to another without traversing the intervening space, represented in modern atomic theory (along with other strange behaviors). He also wanted to 'unify' all the different types of forces we've discovered – magnetism, electricity, gravity, the strong and weak nuclear forces (and whatever new forces were needed to explain things) into one.

So let's say you've read about his failed efforts, then began thinking about it, and then stumbled upon an explanation that, at first, and at least to you, seemed to have solved his elusive theory.

Let's say that you've determined that there really was no significant difference between stars attracting one another (swirling in a galaxy or in binary star systems, or even galaxies clusters), with that of molecules and atoms and subatomic particles (a neutrino is only 1/10,000th the size of an electron, for example) and how they attract each other.

So what you 'realized' is that the significant 'difference' between atoms and stars, besides size, is "time-speed", or the rate at which things happen. For example let's say it takes the average star (our star, for example) 240,000 years (one 'cosmic year') to complete a revolution around the core of the galaxy, while, on an atomic scale, a 'star' (or a subatomic particle) makes it's 'revolution' (around whatever central core it is revolving around) much faster, on a time scale of fractions of a billionth of a second (let's say a scintillionth).

You've deduced this 'time-speed' factor (to which there must be a yet-undiscovered mathematical equation that will explain the relation of time-speed to size, even factoring in force, if it is true) by observing the physical universe – you've observed that things which are smaller tend to 'operate' faster – smaller dog's legs move faster than larger dog's legs, the hearts of birds beat faster than those of elephants, smaller things make higher-pitched sounds (which are 'faster' frequencies), and so on. Further (and this is where solving the unification theory comes in), you observe forces as they affect large objects and small objects – notably static electricity (static 'cling') – where the smaller an object is, the more it tends to 'cling' to you. You extend this size reduction/force increase all the way down to the atomic scale, and you realize that the force of the 'static-cling' of an atom to another atom must be astronomical in relation to their sizes – in other words, theirs becomes the 'glue' that holds objects in our 'macro-universe' together.

Now comes the fun part – 'testing' your theory, your proposed 'model' of the universe (and this is why there are 'different forces' – magnetism, electricity, gravity, strong and weak nuclear - because each could not explain everything on its own, and in experimentation, they all act independently of one another). Let's see if your 'static-cling' unification theory holds up under scrutiny...

Your first question would be, "Why then doesn't everything just come together into a big glob, if everything is attracted to everything else? (just what a universe would do in becoming a singularity again, as a prequel to its next 'big bang') (and perhaps the answer is 'momentum' - the universe is still in a state of counter movement and has not 'settled' yet).

Your second counter-argument comes from quantum mechanics – how can a 'tiny galaxy' emit 'quantum's' of energy? You would expect it to emit a steady stream, with the magnitude varying with heat and pressure.

Another question is why do atoms last so long in their particular forms? We have oxygen 'atoms' and 'hydrogen' atoms that seem to be able to last as long as the universe. Why aren't they constantly changing in form if all they are merely based on are randomly spaced spinning micro-galaxies that are driven by the same static-cling force being proposed?

What about static-cling vs. gravity, for example? We know that they act independently of one another, just as magnetism acts independently, and, supposedly the strong and weak nuclear forces... or do they? Are they actually different, or just different magnitudes (and thus variations) of the same force, and all we are talking about are different time speeds?

Let's suppose you've successfully argued that they are different magnitudes of the same force, and that 'static-cling' is a good representation of it, and that the smaller an object is, the more 'static-cling' it will have, and the more difficult it will be to get it off of you. In this scenario, imagine trying to flick an atom off of your body. You would need to use great force – that which could vaporize you, in fact, to get it off – it is clinging to you with that much 'cling', you would, in effect, need the force required to 'split an atom'. Nuclear radiation would do the trick, or standing inside a particle accelerator...

So, let's say your argument of 'nested' universes (as opposed to parallel universes of the same size) has held up to all questions that have arisen and can account for all observations made in nature. Now what? So we see the universe 'as it is'. So what? What can we do with such knowledge now? Well, 'now' we can't do much because we are still quite stupid! Be patient! At least we can see what 'paths' we must take to navigate through it all, and that is always a beginning...

Concerning such a 'path' and our 'navigating' it, we would know that if we wanted to 'visit' an 'sub-atomic universe', we would not only have to reduce our size, but we would have to increase our 'time speed' to match that of the new environment (but that would happen naturally, as I set forth in the equation below), and we would have to be able to overcome the relative forces that would be acting on us – if we don't want to get 'stuck' there...! (We would need a force behind us to propel us into an 'escape velocity', so to speak).

Let's consider the 'time speed' requirement. Now imagine this – in one sugar cube there are a million billion atoms. Now imagine just one of these with its constituent electrons moving at our 'galactic relative rate' – meaning where an electron, like one of our stars, takes 240,000 of its years to make a circuit around its nuclear core. We would not have to change our 'time speed' at all, only our size.

But now imagine just one of these infinitesimally small electrons going so fast, that the particle can travel from here to our moon in less than a second (meaning at the speed of light) – proportionally, it would be like one of our stars being shot all the way across the universe and the next million billion universes in less than a second. It could be that this is the relative energy required to do what we are suggestion to ourselves - changing our size and 'time-speed'.

Now imagine that one atom, its 'subatomic particles' still moving at the speed of light, 'orbiting' around an infinitesimally small atomic core. It would appear to us that they are everywhere at once – which the current theory of the electron. Now think of the kinetic energy that tiny object has, in other words, what it would do to a ping pong paddle if you tried to 'stop the atom' with it. It probably has enough energy to 'poke a hole' right through the paddle (assuming it would collide with anything, or its trajectory altered by close-encounters with other particles, but more on that later). Yes, it has a lot of energy, and it would be difficult to 'stop' it in its tracks, it is moving so fast – and they say that just in your body alone there is enough potential atomic energy (from your mass) to make thirty large hydrogen bombs (assuming 100% mass-to-energy conversion) (current atomic bombs only change 1%). Thirty hydrogen bombs! Think of it! That is your body's mass-to-energy potential!

So, where was I and where was I going with all of these illustrations...

Oh yes, time speed and size reduction and force resistance (the ability to stay in 'one piece', so to speak, meaning all the relationships of our constituent particles, and, as can now be imagined, we are really talking about reducing THEIR sizes) – the things needed to 'get in sync' with such a tiny system.

To us it seems impossible, but then, we've seen so much already that we suspect it is possible. It could remain science fiction, however. For example they also say that mankind will NEVER reach the edge of our own solar system (meaning the outer edge of the Oort Cloud, which is hypothesized to be just one pitiful light-year from the sun). This means that, no matter how long the species of man lasts, he will NEVER reach the outer edge of the Oort cloud. So maybe 'getting in sync' with an atomic universe will NEVER be attainable after all, and we can only appreciate its possibility (and be thankful they hold us together).

Back to the 'hole in the ping pong paddle' that the electron would make. Well, it really didn't 'make' a hole – there were plenty of 'holes' already there. To the atom, the paddle is mostly empty space – and colliding with another atom was as likely as two stars in our galaxy colliding – very unlikely. Why, they say a neutrino is so small in proportion to the 'empty space' contained in our atoms and molecules that it usually passes right through the ENTIRE EARTH unscathed! It is like a star passing through our galaxy without colliding with another star, and considering the distances between stars, that is actually quite likely. The passing star's course would be minutely altered as it interacted with the gravity of the other stars, but odds are it would continue on its merry way out the other side of the galaxy somewhere.

Do we see any 'evidence' of atoms really being micro galaxies, and why aren't there universes in sizes between us and these atomic universes - in other words, why does matter come in 'quantum sizes'? Do 'galaxy sizes' come only in quantum sizes, just like energy packets? Is there an inverse-square law that describes the size/speed relationship for multi-body system interactions, such as between our galaxy and sub-atomic galaxies?

One thing that the above illustrates is that for every 'answer' we find, it only opens up a Pandora's Box of new questions.

So next time you are licking an ice cream cone, think of all the millions of billions of fantastically fast, super-static-clingy bits of matter that you are consuming, all those hypothetical sub-atomic-scale galaxies, spinning and swirling at the speed of light, and then what will happen to them when they interact with your body's chemistry...

Ah, chemistry. Now where does that fit in...

Happy thinking...!

(a few weeks later...)

The Relativistic Time-Speed/Size/Physical-Speed Equilibrium Equation

So, as Einstein's equations go, the faster you go, the slower you take to do the same action in relation to someone moving at a slower speed (relativistically). This contradicts what we know, or does it? It could explain why atoms and subatomic particles last so long - if they are moving 'at' (or very near) the speed of light, then their time 'stands still' (or very near it).

It also contradicts the notion that the smaller something is, the faster its 'time' goes (meaning it takes less time for 'it' to perform 'one of our' actions, such as scratching one's head); but the more something that small approaches the speed of light, the slower its time goes, meaning 'speed' may play a crucial role in our 'syncing' with a subatomic being.

Which means, according to this theory, that the relatively faster these atomic galaxies go in relation to us (and they seem to move 'near' the speed of light, a speed at which time would be 'nearly' stopped), the slower their relativistic time goes in relation to us (to perform the same action). Without speed involved, a billionth of our second would be a very, very long time for them – meaning they could do a lot of 'living' (perform a lot of actions) in that time-span.

If we are in a time equilibrium – where, if time is a constant, then our physical speeds change to compensate according to our size, and the equation would look like this (I might as well take a stab at the math):

TIME = SPEED/SIZE

which is why 'atomic particles' (what we witness) move at (or near) the speed of light. In other words, to keep relativistic time a constant, smaller objects move faster, according to their reciprocal relationship in relation to time, the constant. So, as our galaxy takes 250,000,000 years to spin once, it only takes a sub-atomic galaxy say a scintillionth of our second to spin once, and, as you can see, we can do a lot of living in 250,000,000 years (as could beings in a sub-atomic galaxy in a scintillionth of one of our seconds).

So in conclusion, there would be a time-speed/size/physical-speed equation to represent this equilibrium between objects of different sizes. One would arrive at the equation through theoretical math (if I already haven't arrived at it by sheer dumb luck above!), which would eventually be verified by direct observation and measurement – that is, comparing the time it takes for objects of the exact same dimensions but of different relativistic sizes to do the exact same action – the smaller object taking LESS time than the similarly-proportioned larger object (if the larger object's time-speed is used as the reference). If time is referenced to the sub-atomic galaxy, then we are performing the action at astronomically slow speeds (relative to them).

Subsequent thinking about the formula, working-in mass (what is a physics equation without mass!) gives a few more variants:

RELATIVE TIME FOR AN ACTION BY TWO PROPORTIONALLY SHAPED MASSES

= MASS 1 / MASS 2

but then we must think about density - our mass can increase without our size increasing. Here we would have to make the applied energy a constant, as well as the proportions of the object (to discount longer arms in the case of head-scratching, for example).

The 31 Million Year Head Scratch

Let's calculate our hypothetical case: Say it takes us (MASS 1) one second to scratch our head. For a 'person' (MASS 2) in our hypothetical sub-atomic sized galaxy, who is, say, one millionth of a billionth of our mass (something like .000000000000001, give or take a dozen zero's), it would take one millionth of a billionth of a second (of our time) for our tiny (relativistically speaking) friend to perform a head scratch. Conversely, it would take us roughly 31 million years in their time to perform our one-second scratch of our head. Also, the duration of their 'galaxy' would be, giving it 20 billion of their years to disperse or coalesce (which would be invisible to us, and regarded as 'energy', the source of new material) roughly 17 seconds in our time.

and

SPEED = TIME* / MASS   (*where TIME is a common unit of reference).

Comparing *TIME = SPEED/SIZE* with SPEED = TIME / MASS (or, rearranged, TIME = SPEED / MASS), we can see that TIME = SPEED / MASS or SIZE, and since DENSITY relates to mass AND size, we can combing the two equations into a related equation:

 TIME = SPEED / DENSITY

and we can also create a variant for relative time:

RELATIVE TIME = DENSITY 1 / DENSITY 2

a few weeks later...

The Shape of Time

Time's 'shape' is composed of 'topographical map' of
speed variations, given a constant momentum; and
you can throw a topographical map of your physical
size in there, too, since size, given an initial density
(meaning not counting the mass picked up due to
speed) , also changes with speed, at least
relativistically.

So let's compare ourselves with our friends living in
one of those micro-galaxies, as if we both began on
our size scale, and try to determine causes and effects.
Let's assume our friend became tiny and we remained
at our present size.

How did they get so tiny? It could have been any one
parameter in any of the equations TIME = SPEED /
DENSITY, TIME = SPEED / SIZE, TIME = SPEED / MASS,
RELATIVE TIME = MASS1 / MASS2, RELATIVE TIME =

DENSITY1 / DENSITY2; and since time is passive, we're left with speed, mass, and density, and out of those, mass and density are passive - we can only really alter our speed, where we 'shrink' not only in the direction we are traveling, but as a whole, which is only possible if we are 'spinning' at a sufficient rate (and with sufficient randomness, which seems impossible given the gyroscopic nature of angular momentum).

So they could have sped-up (the only manipulative variable in this case), reducing their size (relative) while taking on mass, thereby increasing their density.

But, we many ask, why are they 'still here', say within us? Their path must be orbital - meaning circular. If it were linear, we would be saying 'bye-bye' to them.

Now, with larger density, gravity increases, where time is said to slow where gravity is stronger, and time also slows when speeding up, so it follows that time should be much slower for our friend in the micro-galaxy, which counters the observation that smaller things perform tasks faster (leading to the assertion that events happen faster on smaller scales). So we are left with the two possibilities: 1. That the two of us did not begin with equal parameters, or 2. That we did, and all the above effects happened, but in different proportions - where factors speeding time up far outweigh factors slowing down time.

How does this affect the shape of time? One force that the shape of time (varying speeds given a constant momentum) follows are gravity fields, since time slows as gravity increases. So if we wanted to maintain a constant relative speed from our departure point, we would have to increase and decrease our thrust through varying gravity fields - increasing thrust through heavy gravity areas to maintain our initial speed relative to our departure point, and decreasing thrust through lighter gravity zones (using 'fields' 'areas' and 'zones' interchangeably).

Squishing Galaxies

Now imagine trying to 'squish' a smaller galaxy. If the galaxy is on the sub-atomic level, you can't - the force required would be phenomenal - since the force increases as the size difference increases; in other words, you would need an exorbitant amount of force to squish a sub-atomic-sized galaxy.

So where are all the intermediate-sized galaxies? They have all been squished! (or 'consumed'). They have 'interacted' - becoming a part of either our quantum-sized universe or that of the micro-sized; same goes for scales greater than us - everything beyond our cosmic event horizon.

This means quantum size levels for life is determined by 'safe distances' from the next level, where destructive interactions cease, and where all formerly intermediate levels have already been destroyed.

Why did the present ones 'win'? Chance. This leaves us with the existing 'quantum-sized' relationships - where the differences in sizes directly relate to the squish (or 'suck') force required - a smaller-sized galaxy will exist just beyond the squish threshold of matter in the larger galaxy, just as we are just beyond their 'suck' threshold. Mathematically it looks like a probability chart, where the probability of not being squished by objects in a larger galaxy increase to 100% just beyond the squish threshold of objects in the larger galaxy.

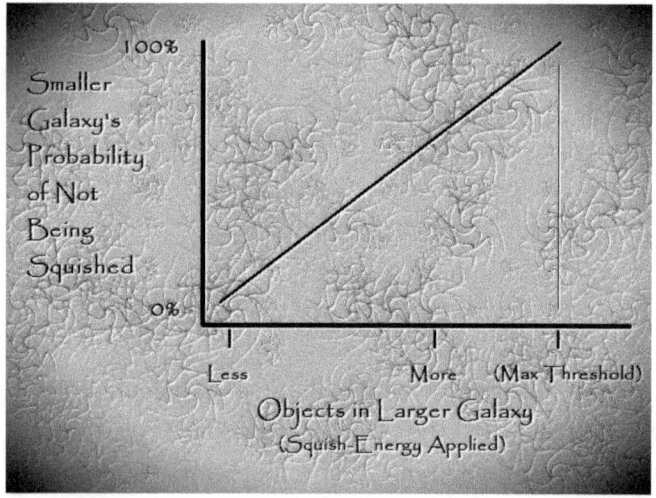

The line represents the smaller galaxy's position within the squish range of the larger galaxy - meaning if the smaller galaxy were beyond the squish limit of the larger galaxy (the smaller galaxy being small enough), it will not be squished further.

The Size Gap Between 'Us' and Sub-Atomic Galaxies

We seem to be made out of atomic particles that are all relatively the same size - we do not see subatomic particles a billion times more massive than others - they all seem to be on the same scale, given a certain small percent of variation. What happened to all the larger particles? Why don't we see quarks six feet in circumference, for example?

This phenomenon would seem to support the notion of 'quantum sizes' - where matter exists at only certain size levels, and where everything in-between was 'consumed' by either the larger state or the smaller state.

Matter and Anti-Matter

As far as 'particles' going in and out of 'existence', it could be merely a micro-universe coming into existence (Big-Bang style, or coalescing) or going out of existence (dispersing, burning-out, or in a Big-Crunch). Incidentally, the aim of life is trying to survive these, perhaps riding them like cosmic surfer-dudes, taking our collection of matter with us to perhaps another quantum size level (if only to share a smile), which implies mutual concern and support between quantum-size separated universes.

The Particle Pair Mystery

It has been observed that, in a super-collider, when one particle is knocked out of an atom and spins away, a mutual particle, somewhere on the other side of the collider, is 'knocked away' and spins in the opposite direction, which is as yet unexplained (whether this really happens or our observational power is not yet fast and powerful enough to see what is really happening). Assuming it is true, one thinks of 'wormholes' (which may not exist, being but a mathematical graphical abstraction).

In our case, two distantly separated micro-universes would be 'connected', and, as we destroy one, another is destroyed, and, as it just happens by chance, just far enough away to be on the other side of the super-collider and observable; and this observation remains to be explained. Perhaps, being connected by a wormhole, when one is annihilated, the wormhole connection is lost, and the other 'snaps' away, like two halves of a rubber band under tension breaking and snapping back in their opposite directions.

Inter-Universe and Extra-Universe Space

As far as wormholes being tunnels to other universes, or to a distant point in the same universe, where two black-hole 'gravitational pits' touch and are bound to one another, you would be crushed at the midpoint trying to travel through it; but let's say you have found a way to survive that - now graphically speaking, it looks like this:

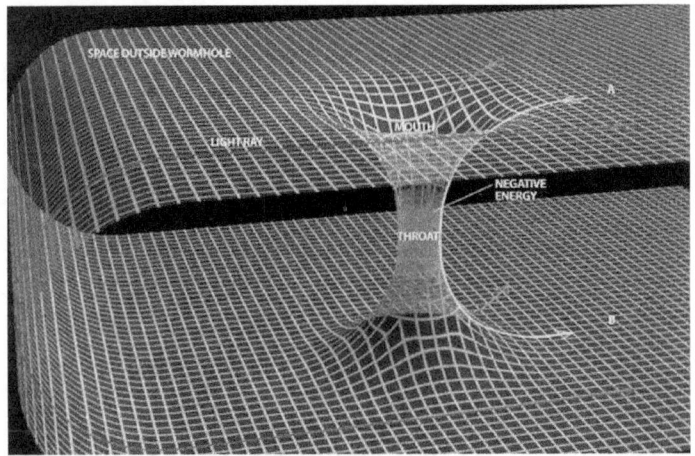

As you can see, you would 'fall' into the 'mouth' and 'throat' of that black hole's gravitational field, but you would need an exorbitant amount of energy to exit out the other side - needing to escape the gravity of that black hole. Now, graphically speaking, you can see that, if you are in the 'throat', and you poke your way

313

out through the 'side' of it, you would be in no-man's land - you would be in an 'inter-universe' space, or 'extra-universe' space (which is where all my lost keys go, no doubt, and an expedition to retrieve them would be in order).

As far as where we fit into this picture in relation to our micro-universes, you can see that we exist primarily in this inter-universe/extra-universe space, as far as they are concerned, in fact, we would embody it.

Contacting Your Friend in a Micro-Universe

Since your friend lives and dies in roughly a trillionth of a billionth of one of our seconds, you would need an exceedingly high electromagnetic frequency within which to encode your message. This bring up the question of which frequency to use, and, considering the well-being of our friend, which wavelength is a harmonic of its galaxy - where, if we send too strong a signal in that frequency, it would vibrate our friend's galaxy apart, shattering it. Imagine that! At any rate, such a frequency is most likely still beyond our present technology, and maybe beyond our friends - for detecting a wavelength that spans an entire galaxy is difficult to detect!

There are two other problems. It is possible that not all 'galaxies' (universes) will produce 'life', or at least advanced, thinking (sentient) lifeforms. The probability could be great, though it could be very small. Let's say it is a billion to one. That still leaves a LOT of atomic galaxies on the tip of your nose.

Another problem is timing. Consider our place in our universe. We've been 'sentient' for only a billionth of our universe's time span. If we can perpetuate our survival for some time to come, we would increase the chances of the next larger quantum-sized life 'making contact' with us; and so we have the same challenge with finding and making contact with our atomic friends as they evolve and struggle to withstand all that their universes throw at them.

In summary, it is quite a challenge for all of life to 'connect' considering the space/time hurdles - in space they are relative quantum magnitudes and relative locations, and in time they are relative rates and timing (meaning 'luck' assuming a random, chaotic system of physicalities).

So we've looked up - to ponder the larger quantum-sized galaxies, and we've looked down, to ponder the smaller quantum-sized galaxies, and we've looked out, to ponder other life on our universe; now we can look in - to the atoms within us. Imagine modifying our biological senses to the extent where we could detect and communicate with the denizens of any micro-galaxy within us - you would suspect this to be a telepathic endeavor; and what would you say to them? "Hello there, you are inside my body." Fanciful thinking indeed.

Quantum Boundaries

These can be defined by the limits of that which affects you in a practical sense. If sub-sub-sub-sub-sub atomic particles do not really factor into anything that affects us, then such particles are beyond our quantum boundary.

One possibility is that we do not share the same quantum boundaries with our neighbors (those that are one level smaller and those one level larger) - there may be gaps between our boundaries, small or significant.

I would venture that they are small - as anything that mutually interferes would have been eliminated, leaving no gap - because the universe never does any more work than is absolutely necessary to achieve equilibrium.

Technological Progress in Terms of Orders of Magnitude (Zero's)

Technological progress can be measured in orders of magnitude - "zero's" - in other words - how many zero's closer are we to a goal, and in this case it would be detecting the denizens of our neighboring quantum-sized galaxies (both the next step larger and the next step smaller). As an example, say we wish to detect a signal from the denizens living in a next quantum-level size smaller galaxy. Let's say their wavelength, as slow as they can get it, turns out to be our wavelength of a billionth of a billionth of a billionth of a second, or a wavelength that is around 0.00000000000000000001 angstroms in length, and let's say our current technology can only detect a wavelength down to .0001 angstrom. You can see that 'progress' is simply a matter of reaching the next order of magnitude - adding to/subtracting from (as the case may be) some of those zero's. In the case of detecting our minute friends, for example, our instruments are 16 zeros (orders of magnitude) away in range and sensitivity.

The Other Math in This Scenario

Let's say that the life of the universe one quantum-size level down lasts 20 billion years, and let's say it is 20 of our seconds, and let's say that life enlightened enough to suspect our existence exists for only 1 second of that duration; then we are limited to one second to try detect them (via the signals they send us).

Now let's consider frequencies and intensities. Let's say they are a billion, billion, billion times removed from us - the frequencies faster and the intensities weaker, and this though they are doing the best that they can - transmitting as low a frequency as they can at the strongest signal they can. If they were really advanced and powerful, they could get an entire galaxy to pulsate at one pulse every few million years, and at a strong enough signal that we could detect it. This means that our most sensitive hyper-microwave receivers needs to be millions of billions of times more sensitive to detect them, and the frequency it must detect will be a million billion times higher than our equipment's current range, which means our current technology is a factor of around 15 zeros away from achieving those levels. We can expect that we will slowly edge closer, maybe shaving off a zero every hundred years as technology advances. In numbers, it would look something like this:

Weakest Signal that Our Instruments Can Detect: "1"

Strength of Their Signal: 1 billionth of 1.

Difference: A factor of nine zeros (using the exponent of the power of 10), so we are 'nine zeros' away from possible success.

The same holds true for frequency:

Highest Frequency Our Instruments Can Detect: "1"

Their Frequency: 1 billion times higher

Difference: a factor of nine zeros (10 to the 9th power).

The above assumes the relationship is linear, which it may not be, since our universe loves squares in its equations.

The same math would hold true for our biological senses if we were trying to detect the denizens in the atoms within us.

The Lifespan of Our Universe

Now, if we are a micro-universe in someone else's space-time, what they do may affect the life span of our universe. Let's say we are embedded in their cake batter. Now, when they place that batter in the oven, the heat may affect our universe's life span (I would assume the worst and expect it would reduce it. and plot my goals and future actions along those lines).

So in summary, the longer a universe lasts, the higher the chances of unexpected transformations.

Unforeseen Larger Structure Life Functions

Just as atoms and molecules come together to form a larger structure that then performs an unforeseen life function (say a cell), and just as those cells come together themselves to form a larger structure that itself performs an unforeseen life function (say a liver), so too can galaxies and universes (within the same quantum size) come together to create a larger structure, that then begins to perform an unforeseen life function on a larger plane. We cannot perceive it due to the extreme time-speed differences between quantum-size planes, which may be due to a limit in energy expenditure over time - in which, if that life function (say a scratching of an enormous head) occurred any faster, it would annihilate us and itself due to too much energy being expended too quickly. So the action occurs at a 'safe' speed, within the limits, which may be a constant, and which is much too slow for us to observe (the one-second head scratch of the enormous other-quantum-size plane head lasting 32 million of our years, for example), which wouldn't happen - for it would need to use more energy than is available to it.

Size Ergo Speed, or Speed, Ergo Size?

As an object approaches the speed of light, its mass increases toward infinity and its size approaches the infinitely small. So that means our tiny friends may have started out our size, but now that they're moving near the speed of light, they are infinitely small to us; and since the mass of our friends is nearly infinitely amplified at that speed, it means our friends are also actually in a much smaller state than we realize, since we are detecting their nearly infinitely amplified mass. It also means that if we slow our friends down to 'our speed' their mass will return back to 'normal', and they will balloon back up to our size. The problem may be that they are stuck at that speed and size; perhaps leaps in quantum speeds are only one way, or maybe he only knows how to jump to a higher quantum speed. At any rate, we do not see people ballooning out of 'thin air' right before our eyes (as they quantum leap back to 'our speed', or 'disappearing (quantum leaping to a faster speed) – meaning 'we' do not know how to do it yet, and 'they' having bothered to travel to our quantum speed level.

So when I say there are quantum sizes, and that speed is affected by sized (where size is the cause and speed is the effect), it may be that speed be the cause and size the effect.

The "Infinite Number of You's"

Just to expand your imagination, as for the infinite number of "you's" out there (which is a certainty if infinity exists), they now not only include all the "you's" that are your size, but also macro and micro "you's" existing in those macro and micro galaxies.

Now, just a parting note here - more important than being right, I'm demonstrating the spirit of exploration and discovery... the joy of being right can wait... (until I am truly right!)

On the Quality of Our Existence

On a related note, a space pundit recently suggested that advanced life forms will have found a way to slow their existence, thereby 'living longer'. Now to this I'd say it is irrelevant how 'long' you live relative to eternity – we are merely sampling it at a certain rate. This brings up the issue of 'quality'. Length along does not assure quality, but the events within your time frame, and not even the number of events – how many you can squeeze into your existence, but the quality of those events. The same can be said for size – in space – it is not the number of artifacts you create, but their quality.

What Defines Our Quantum Size?

That would be the size of our most basic building blocks (atoms). What determined the size of our atoms? That would be the size of our Big Bang. So, theoretically, to determine the size of our Big Bang, we could trace it back via the size of the building blocks that it created. This means that there may be (it would be a certainty if infinity and eternity existed) other Big Bang created universes with different sized 'atoms', creating different sized beings and objects.

This can be extended to the properties and laws of our universe - being custom-designed by the size and nature of our Big Bang, and which also could be used to trace back the size and nature of our Big Bang.

Concerning eternity and infinity, we are merely 'sampling' each in terms of speed and size, and it doesn't really matter which speed and size we are (since eternity and infinity are boundless), unless we are trying to interact with life forms in different quantum speed/size states.

Incidentally, the length of time does not define the quality of our limited existence, nor does the number of events one can squeeze into one's time frame - it is rather the quality of the events that matters.

What Kind of Life Forms Would 'They' Be?

Consider ourselves. We are a trial-and-error experiment in size and brain capacity. Size-wise, we may be the current dinosaurs - too large. Consider how much brain matter we actually use to be technologically intelligent, and that is the ideal size, the ideal size being 'minimum'. Let's be generous and say it takes a brain half our size to achieve such a technological level as to communicate across a quantum barrier, then that would be their size.

Now our bacteria are too small and live too short a lifespan to become technology-oriented. Larger animals are already on the verge of extinction from small entities (us). So our quantum-sized friends are probably the size, relatively speaking, of half of our brains, that being the minimum necessary to achieve such a technological level; and there may be large-armed versions for performing the necessary heavy labor!

Observations that Indicate None of This is True

Let's consider the splitting of atoms into smaller particles. You would think that, if we were tearing little galaxies apart (and destroying all life within them, which brings up the possibility of atom rights activists), the component parts would be random in size, and not in quantum sizes, as in the specific-sized subatomic particles that are seen. This would argue against little galaxies, at least on an atomic quantum level. The question would be, where do the particles stop and the galaxies begin (going from standard sub-sizes to random), which means, when a subatomic particle is split, are the component parts uniform (particles) or random (where we've reached the next quantum level of life). To split smaller and smaller particles, we need more and more power - and we are talking about power levels with a lot of zeros, where we still have a lot of zeroes to go to get there.

Just a note - though this model may not be true, its nagging possibility should increase the science community's sensitivity to ethics. For example, when a future subatomic particle is split, the scientists will wince, hoping the component parts will be uniform in size (indicating another level of sub-particles) and not random (indicating that the next quantum-sized galaxy has been reached, and that it has been torn apart, and along with the possible destruction of the galaxy, the destruction of all life within it), meaning science will proceed more cautiously, and with more regard for what they are about to have an effect on, rather than proceeding as blind bulls in china shops.

Back to our universe. You might wonder if our universe was in a supercollider, and our galaxy was 'split' like an atom in a super-collider, how long would it take, and would we notice it? If one second of theirs is 32 million years of ours, and it only takes a billionth of a second, then it would only take 11 days. Would we notice it? We sure would! The stars. or, on a broader level, the galaxies, would swiftly and alarmingly begin to swiftly move and rearrange themselves. Would it mean our demise? Maybe, maybe not - it depends on our trajectory and what is in our path, and how interacted we were with what is 'out there' on our quantum size level. We can count on a lot of nothingness (empty space), so there is hope that we would not fatally collide with anything, or lose any critical interactions with distant stars and galaxies.

Another question would be what about atoms at the center of stars - aren't their environments too harsh for micro-life? Answer: If so, then it can be seen that not every 'universe' is conducive to life, in fact the chances may be very small - perhaps nil for those atoms; but then again, life has surprised us many times already in existing in seemingly impossible places and environs.

Another question would be that if this 'quantum size' model were true, and if Eternity does exist, then you would expect a lifeform on some quantum plane to have discovered how to communicate across all the quantum barriers, and we would already have been contacted and made aware of all such facts. An explanation as to why we do not is that 'success' may be cyclical - for example, 'found' when 'good' predominates, and then 'lost' when 'evil' predominates. Right now, since we do not know anything, one may conclude that we are in some kind of evil era where information is withheld by some evil tyrant.

 Now for the biggest argument against all this - the lifespan of an atom. These 'micro-galaxies', supposedly our 'atoms', only last around seventeen seconds (or some very brief timespan), and yet atoms are said to last as long as our universe - trillions of years (it is what keeps us 'together' - what keeps oxygen 'oxygen', us 'us', and stars 'stars', for example). If our atoms were such short-lived 'galaxies', all of our matter would 'fall apart' in seventeen seconds. So the answer must be that our atoms are mere particles after all, and are not such short-lived micro-galaxies, and, if it is not reality, it is nothing we'd want to base a new world philosophy on.

There are alternate possibilities (maybe endless) - that the next quantum level down (for intelligent life, which would its own atomic-sized particles with which to be made of) is contained in our subatomic particles, or something along that line, but that makes them even more distant, and even less likely that we will ever detect them. So, for now, we are, in practicality, stuck with, and should only be concerned with, our own universe, within whatever system is out there.

Science vs. Religion

What if none of this turns out to be true, yet we want to believe it anyway? Then it becomes a 'religion', or a philosophy. Ideally, you would want any misery-alleviating philosophy to be based on reality, meaning science. Failing that, you are left with religion - or pure philosophy, until a more accurate way of thinking is developed.

 What this means is, since the theory (of quantum sized units for universes - meaning matter, energy, and any life within) was not derived through mathematical discovery or direct observation (only extended, hypothetical observations and the most cursory math), it is scientifically weak (being untestable)(at least so far, since it is beyond our instruments) (but that doesn't mean the experiments aren't there), so there is a very good chance that none of it is true; yet it does give us a nice frame of mind within which to exist, meaning people have the option of using it, like a tool, to make their lives more meaningful, and perhaps bearable, and to provide a new framework for defining 'good and evil' (the 'conducive and the non-conducive' toward a goal - this goal being trying to detect and communicate with such life on other quantum-size levels).

What about current religions? Any science can easily fit under the umbrella of any pan-religion or pan-mysticism just by making claims (that 'whatever is' is simply a part your Supreme Being's plans and designs).

Why Detect and Communicate?

Why should we try to detect, and then communicate, with life on another quantum plane? I can immediately think of two reasons: 1.) They may know things that we do not know and have answers we do not have; and 2.) We, being heroic in nature, may find we are needed to 'save' them.

What would their 'signal' be like? It would consist of repeating pattern transmissions, something like Morse Code, something like the SOS signal (three short bursts, three long bursts, three short bursts) which would clearly distinguish it from nature. The patter may be shorter and simpler, since, for them, the duration of each pulse may be millions of years, and the intensity requiring the energy equivalent to many supernovae.

A Philosophical Issue

A question we can ask is, "Where are we in the scheme of the universe one quantum-size up, and how can we make a positive impact?" The answer to the first part of the question may very well be that we are within one subatomic particle of that universe. What impact we can have from that perspective is the challenge - should we move our universe right or left, for example, whether we are lucky enough to be at the core of a protein, or stuck at the center of a star - what could we do in our billionth of a second existence and on such a tiny scale?

Near-Penultimate Thought (The Inevitable) - Ego Daydreams

So there I am, accepting my Nobel Prize for all of this. Several problems immediately arise - first, they aren't sure what field to put this in, so they consider a Nobel Prize for 'Theory', or maybe for just plain 'Thinking' (and inspiring others to think), most likely it would be in Cosmological Modeling. Second, since I didn't do the 'heavy math', it would have to be a 'Light' version!

So there I am, accepting my Nobel Prize 'Light' in the field of Cosmological Theory, while real astrophysicists take the reins and dive into the heavy math, to try and prove or disprove the theory. They would lean toward proving it, I imagine, for who wins a Nobel Prize 'disproving' something? The danger in that is that it would lead to bias, and the resulting erroneous conclusions.

If I am Proven Wrong

If all this is just the "fantastic creation of a speculative philosopher" (not my line) then I will solemnly bury the notion with due parting fondness and sentimentality (and a fitting eulogy), and move on - after all, I am into reality, and not into creating myths; but how can I be proven wrong, given the infinite bounds of possible size within the infinite bounds of space, in which we have an infinite amount of time to perform one action? Young physicists-to-be (school boys) tend to think that "all there is" is what science can presently 'see' (and we can forgive them for that, even in the face of the obvious); but it is healthier (in terms of survival) to realize that whatever we do today, and whatever we know, and especially whatever technology we create, will be, by tomorrow's standards, primitive, clunky, weak, slow, silly, awkward, and ugly; and, regarding many current scientific theories, just plain wrong...!

I suppose the questions to ask are, "If this is true, what does it imply?" (Meaning "What can we do with it?" - for example atomic theory gave us nuclear power and quantum mechanics gave us semiconductors and lasers, among other things).

Do We Need to Pursue Micro-Galaxies Today?

No - for two reasons: 1. they have not posed a threat, and probably will not in the foreseeable future; and 2. no matter what potential benefits we would gain by interacting with such beings, they are still far beyond our technology to detect (assuming we would need the aid of technology as an intermediary in light of our biological senses having become less useful as we face more difficult problems).

Answering the Question "Universe or Multiverse"

Multiverse. Each universe contains a specific amount of matter/energy, and they are all separated by distance, or time, or size (size refers to what I will dub "nested universes" - which all share a common point in space and time. and they are in quantum sizes where they no longer 'interfere' with one another) (and different sizes mean different speeds, too), or they are separated by combinations thereof (for example, distance and time, or size and distance, or size and time and distance). Seems pretty clear and simple to me. Now consider the absurd statement that the Big Bang's singularity contained all matter (and the even more absurd statement that there was no "space" before the Big Bang) - because given a boundless infinity, there is a boundless "everything", far beyond our Big Bang Universe, which brings up the notion of "nothing" - which could never have existed since we have "something", and you cannot create something from nothing, nor can you ever achieve "nothing", since no matter where you try to hide your "something" (in order to achieve "nothing"), it will be "somewhere" (and in some form). Just sayin'; and, just for the record, infinity (and eternity) (and "everything") as "wholes" do not exist to us, nor would we exist to them.

The Danger in "Believing"

Say you want to believe in all of this, and then you see something that violates it - that breaks its rules, that 'should not happen': You will in effect be 'blind' to it, you will not believe it, perhaps long enough for 'it' to bite you. That is not a good thing. One thing I can think of is some entity appearing as if shredding through space from a parallel, superimposed universe when such superimposition (by 'undetectable' universes within the same quantum size) (I say 'undetectable' because two universes can come to overlap as they expand, but we would be able to detect it, all physical matter on our quantum plane being compatible, or at least would interact until they mix or one 'consumes' the other) should not be possible; meaning if it does, then that instantly presents a different reality than that which our model, which we assumed was correct, which common sense agrees with, represents, invalidating our model.

A Rare Catch

Just as we may travel from galaxy to galaxy, so too may our minute friends - only for them they would be traveling between our atoms. So we may not only detect them within atoms, but between them, which would be a rare catch indeed. This brings up the question of whether they can see one another's 'galaxies' (our adjacent atoms).

Mathematically Defined vs Reality

It is easy to see that many of the properties of such nested universes, galaxies, and denizens can be mathematically calculated, but does this prove that they actually exist? No - 'proof' comes with empirical evidence.

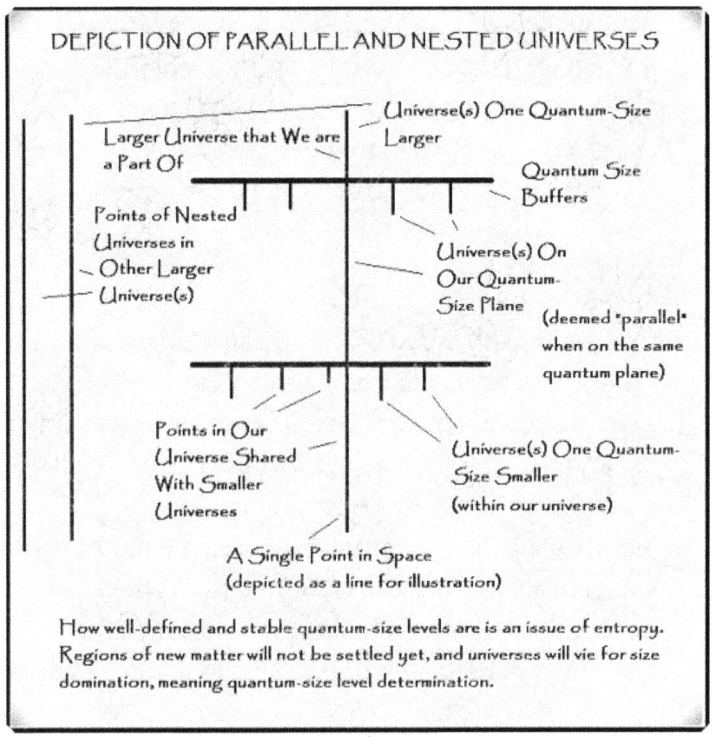

DEPICTION OF PARALLEL AND NESTED UNIVERSES

Universe(s) One Quantum-Size Larger

Larger Universe that We are a Part Of

Quantum Size Buffers

Points of Nested Universes in Other Larger Universe(s)

Universe(s) On Our Quantum-Size Plane

(deemed "parallel" when on the same quantum plane)

Points in Our Universe Shared With Smaller Universes

Universe(s) One Quantum-Size Smaller

(within our universe)

A Single Point in Space
(depicted as a line for illustration)

How well-defined and stable quantum-size levels are is an issue of entropy. Regions of new matter will not be settled yet, and universes will vie for size domination, meaning quantum-size level determination.

Distances Between Same-Quantum Sized Universes

Say we limit our definition of 'universe' to the matter and energy originally contained in its originating Big-Bang (so 'Big-Bang Universes'), and say we assume that our local Big-Bang was merely one of many. Then the question arises, how far apart are we from other Big-Bang universes (which exist on the same quantum-size plane as ours), provided they have not completely dissipated yet)?

We can use what we know already about stars and their relative sizes and distances (the distances between each star being far, far greater than the size of the stars). We could use the size-distance relationship between galaxies (if it is different from stars - and if it isn't, then we have discovered a general size-distance rule).

Let's get back to our neighboring same-quantum-size-plane universe. It could very well be slightly different as far as a quantum size goes - meaning it would not 'fit' into our local scheme of quantum sizes (that which our universe has settled into) - it may have intermediate sizes for atoms, meaning if that universe and our collided, there may be 'quantum-size conflicts' - energy/matter interactions that would annihilate some sizes during the process of creating a new 'stable system' of non-interacting quantum sizes.

Let's extend this to our universe, and to all the micro-universes within it. What this would mean is that our micro-universes would not necessarily be compatible with each other, or the same 'sizes', atomically speaking. This would make them all different sizes, and we do not see that in nature - all electrons, for example, are the same relative size, meaning 'electrons' themselves may not be micro-universes proper, since there would be a great number of size variations, made possible by the distances between them, rendering them non-interactive, where they would not destroy one another as a new size equilibrium was settled into.

It could be that there WERE once a great number of size variations in universe frameworks (such as the size of the atoms in them), but things HAVE settled into a 'universal' equilibrium on that quantum-size plane, but that would entail all atoms (micro-universes) to have 'interacted', which doesn't seem possible, given the distances between atoms and the vast space of our universe- but then, given eternity, perhaps all this interaction has had 'time' to occur.

On Universe Dissipation

It appears that our visible universe is expanding, as if dissipating into the void, like a cloud of gas that originated in an explosion. The question now becomes who are our non-dissipated neighboring universes, and are we 'collection' any of that dissipated matter and energy (or just energy, if all the matter has already dissipated into energy).

Another question arises - how can we survive our own dissipation, or will we dissipate, and our energy merely be 'collected' by other systems to be incorporated into, or form, another universe? (and this is an example of philosophy giving investigative science direction).

On the Intermediary 'Safe Zones'

Questions will arise as to how the 'safe zones' between the nested universes are created and maintained (the study of which may lead to answers that we would expect would only arise from directly studying the universes themselves).

Now here, as with all hypotheses, if every aspect cannot be mathematically modeled, then as a theory goes, it is a cause for despair for that theorist for that theory.

Happy mutual 'waving' with a person in an atomic galaxy (and to the infinite number of "you's" out there)!